P9-BYV-044

THE CANON

BOOKS BY NATALIE ANGIER

NATURAL OBSESSIONS

Striving to Unlock the Deepest Secrets of the Cancer Cell

THE BEAUTY OF THE BEASTLY

New Views on the Nature of Life

WOMAN

An Intimate Geography

THE CANON

A Whirligig Tour of the Beautiful Basics of Science

THE CANON

A Whirligig Tour of the Beautiful Basics of Science

Natalie Angier

HOUGHTON MIFFLIN COMPANY
Boston • New York 2007

Visit our Web site: www.houghtonmifflinbooks.com.

Library of Congress Cataloging-in-Publication Data
Angier, Natalie.
The canon : a whirligig tour of the beautiful basics of science / Natalie Angier.
p. cm.
Includes bibliographical references and index.
ISBN-13: 978-0-618-24295-5
ISBN-10: 0-618-24295-3
1. Science—Popular works. I. Title.
Q162.A59 2007
500—dc22 2006026871

Printed in the United States of America

Book design by Robert Overholtzer

MP 10 9 8 7 6 5 4 3 2 1

FOR RICK,

my one in 6.5×10^9

Contents

THE CANON

Introduction

Sisyphus Sings with a Ying

WHEN THE SECOND of her two children turned thirteen, my sister decided that it finally was time to let their membership lapse in two familiar family haunts: the science museum and the zoo. These were kiddie places, she told me. Her children now had more mature tastes. They liked refined forms of entertainment — art museums, the theater, ballet. Isn't that something? My sister's children's bodies were lengthening, and so were their attention spans. They could sit for hours at a performance of *Macbeth* without so much as checking the seat bottom for fossilized wads of gum. No more of this mad pinball pinging from one hands-on science exhibit to the next, pounding on knobs to make artificial earthquakes, or cranking gears to see Newton's laws in motion, or something like that; who bothers to read the explanatory placards anyway? And, oops, hmm, hey, Mom, this thing seems to have stopped working! No more aping the gorillas or arguing over the structural basis of a polar bear's white coat or wondering about the weird goatee of drool gathering on the dromedary's chin. Sigh. How winged are the slippers of time, how immutably forward point their dainty steel-tipped toe boxes. And how common is this middle-class rite of passage into adulthood: from mangabeys to Modigliani, *T. rex* to *Oedipus Rex.*

The differential acoustics tell the story. Zoos and museums of science and natural history are loud and bouncy and notably enriched with the upper registers of the audio scale. Theaters and art museums murmur in a courteous baritone, and if your cell phone should bleat out a little Beethoven chime during a performance, and especially should you be so barbaric as to answer it, other members of the audience have been instructed to garrote you with a rolled-up *Playbill.* Science appreciation is for the young, the restless, the Ritalined. It's the holding-pattern fun you have while your gonads are busy ripening, and the day that an exhibit of Matisse vs. Picasso in Paris exerts greater pull than an Omnimax movie about spiders is the debutante's ball for your brain. Here I am! Come and get me! And don't forget your Proust!

Naturally enough, I used the occasion of my sister's revelation about lapsing memberships to scold her. Whaddya talking about, giving up on science just because your kids have pubesced? Are you saying that's it for learning about nature? They know everything they need to know about the universe, the cell, the atom, electromagnetism, geodes, trilobites, chromosomes, and Foucault pendulums, which even Stephen Jay Gould once told me he had trouble understanding? How about those shrewdly coquettish optical illusions that will let you see either a vase or two faces in profile, but never, ever two faces *and* a vase, no matter how hard you concentrate or relax or dart your eyes or squint like Humphrey Bogart or command your perceptual field to stop being so archaically serial and instead learn to multitask? Are your kids really ready to leave these great cosmic challenges and mysteries behind? I demanded. Are *you?*

My voice hit a shrill note, as it does when I'm being self-righteous, and my sister is used to this and replied with her usual shrug of common sense. The membership is expensive, she said, her kids study plenty of science in school, and one of them has talked of becoming a marine biologist. As for her own needs, my sister said, there's always PBS. Why was I taking this so personally?

Because I'm awake, I muttered. Give me a chance, and I'll take the jet stream personally.

My bristletail notwithstanding, I couldn't fault my sister for deciding to sever one of the few connections she had to the domain of human affairs designated Science. Good though the Oregon Museum of Science and Industry may be, it is undeniably geared toward visitors young enough to appreciate such offerings as the wildly popular "Grossology" show, a tour through the wacky world of bodily fluids and functions.

Childhood, then, is the one time of life when all members of an age cohort are expected to appreciate science. Once junior high school begins, so too does the great winnowing, the relentless tweezing away of feather, fur, fun, the hilarity of the digestive tract, until science becomes the forbidding province of a small priesthood — and a poorly dressed one at that. A delight in "Grossology" gives way to a dread of grossness. In this country, adolescent science lovers tend to be fewer in number than they are in tedious nicknames: they are geeks, nerds, egg-heads, pointy-heads, brainiacs, lab rats, the recently coined aspies (for Asperger's syndrome); and, hell, why not "peeps" (pocket protectors) or "dogs" (duct tape on glasses) or "losers" (last ones selected for every sport)? Nonscience teenagers, on the other hand, are known as "teenagers," except among themselves, in which case, regardless of gender, they go by an elaboration on "guys" — as in "you guys," "hey, guys" or "hey, you guys." The you-guys generally have no trouble distinguishing themselves from geeks bearing beakers; but should any questions arise, a teenager will hasten to assert his or her unequivocal guyness, as I learned while walking behind two girls recently who looked to be about sixteen years old.

Girl A asked Girl B what her mother did for a living.

"Oh, she works in Bethesda, at the NIH," said Girl B, referring to the National Institutes of Health. "She's a scientist."

"Huh," said Girl A. I waited for her to add something like "Wow, that's awesome!" or "Sweet!" or "Kewl!" or "Schnitzel with noodles!" and maybe ask what sort of science this extraordinary mother studied. Instead, after a moment or two, Girl A said, "I hate science."

"Yeah, well, you can't, like, pick your parents," said Girl B, giving her beige hair a quick, contemptuous flip. "Anyway, what are you guys doing this weekend?"

As youth flowers into maturity, the barrier between nerd and herd grows taller and thicker and begins to sprout thorns. Soon it seems nearly unbreachable. When my hairstylist told me he was planning to visit Puerto Rico, where I'd been the previous summer, and I recommended that he visit the Arecibo radio telescope on the northwestern side of the island, he looked at me as though I'd suggested he stop by a manufacturer of laundry detergent. "Why on earth would I want to do *that*?" he asked.

"Because it's one of the biggest telescopes in the world, it's open to the public, and it's beautiful and fascinating and looks like a giant mirrored candy dish from the 1960s lodged in the side of a cliff?" I said.

"Huh," he said, taking a rather large snip of hair from my bangs.

"Because it has a great science museum to go with it, and you'll learn a lot about the cosmos?"

"I'm not one of those techie types, you know," he said. Snip snip snip snip snip.

"Because it was featured in the movie *Contact,* with Jodie Foster?" I groped frantically.

The steel piranhas could not be stilled. "I've never been a big Jodie Foster fan," he said. "But I'll take it under advisement."

"Hi, honey!" my husband said when I got home. "Where did you put your hair?"

In truth, I pull it out myself just fine, all the time. How could it be otherwise? I am a science writer. I've been one for decades, for my entire career, and I admit it: I love science. I started loving it in childhood, during trips to the American Museum of Natural History, and then I temporarily misplaced that love when I went to a tiny high school in New Buffalo, Michigan, where the faculty was so strapped for money that one person was expected to teach biology, chemistry, and history before dashing off for his real job as the football coach. The overstretched fellow never lost his sense of humor, though. One morning, as I approached his desk to present him with my biology project, a collection of some two dozen insects pinned to cardboard, I noticed that the praying mantis, the scarab beetle, and the hawk moth were not quite dead, were in fact wriggling around desperately on their stakes. I screamed a girlish stream of obscenities and dropped the whole thing on the floor. My teacher grinned at me, his eyes merrily bug-eyed, and said he couldn't *wait* until it was time for me to dissect the baby pig.

In college I rediscovered my old flame, science, and it was still blazing Bunsen burner blue. I took many science courses, even as I continued to think of myself primarily as a writer, and even as my fellow writers wondered why I bothered with all the physics, calculus, computers, astronomy, and paleontology. I wondered myself, for I was hardly a natural in the laboratory. I studied, I hammered, I nattered, I plucked out my hairs, but I kept at it.

"Well, aren't you a little C. P. Snow White and the Two Cultures," said a friend. "What's your point with these intellectual hybridization experiments, anyway?"

"I don't know," I said. "I like science. I trust it. It makes me feel optimistic. It adds rigor to my life."

He asked why I didn't just become a scientist. I told him I didn't want to ruin a beautiful affair by getting married. Besides, I wouldn't be a very good scientist, and I knew it.

So you'll be a professional dilettante, he said.

Close enough. I became a science writer.

So now, at last, I come to the muscle of the matter, or is it the gristle, or the wishbone, the skin and pope's nose? I have been a science writer for a quarter of a century, and I love science, but I have also learned and learned and not forgotten but have nevertheless been forced to relearn just how unintegrated science is into the rest of human affairs, how stubbornly apart from the world it remains, and how persistent is the image of the rare nerd, the idea that an appreciation of science is something to be outgrown by all but those with, oddly enough, overgrown brains. Here is a line I have heard many times through the years, whenever I've mentioned to somebody what I do for a living: "Science writing? I haven't followed science since I flunked high school chemistry." (Or, a close second, ". . . since I flunked high school physics.") Jacqueline Barton, a chemistry professor at the California Institute of Technology, has also heard these lines, and she has expressed her wry amusement at the staggering numbers of people who, by their own account, were not merely mediocre chemistry students, but undiluted failures. Even years of grade inflation cannot dislodge the F as the modal grade in the nation's chemistry consciousness.

Science writing, too, has remained a kind of literary and journalistic ghetto, set apart either physically, as it is in the weekly science section of the *New York Times,* or situationally, as it is by being ignored in most places, most of the time, no matter how high the brow. Ignored by *Harper's,* ignored by the *Atlantic,* ignored by, yes, *The New Yorker,* ignored by the upscale cyberzines like *Salon* despite the presumably parageek nature of their audience. I've seen reader surveys showing that, of all the weekly pull-out sections in the *New York Times,* the most popular is "Science Times," which runs on Tuesdays. Yet I also know, because I have been told by kindhearted friends and relations, that many people discard the whole section up front and unthumbed. Some of those preemptive ejectors even work for the *New York Times.* Several years ago, when the woman who was then the science editor of the *New York Times* asked the man who was then the chief editor of the entire paper to please, please, give the science staff some words of appreciation for all their good work, the chief editor sent a memo assuring the staff how much he looked forward to "Science Times" . . . every Wednesday. When I first started writing for the newspaper, and I introduced myself as a science reporter to the columnist William Safire, he said, "So I would be likely to read you on Thursdays, right?" Harold Varmus, a Nobel laureate, told me I should have replied, "Sure, Bill, if you read the paper forty-eight hours late."

Oy, it hurts! How could it not? Nobody wants to feel irrelevant or marginal. Nobody wants to feel that she's failed, unless she's in a high school chemistry class, in which case everybody does. Yet I'll admit it. I feel that I've failed any time I hear somebody say, Who cares, or Who knows, or I just don't get it. When a character on the otherwise richly drawn HBO series *Six Feet Under* announces that she's planning to take a course in "biogenetics" and her boyfriend replies, Bo-o-ring. Why on earth are you doing that? I take it personally. Wait a minute! Hasn't the guy heard that we're living in the Golden Age of Biology? Would he have found Periclean Athens bo-o-ring too? When my father-in-law finishes reading something I've written about genes and cancer cells and says he found it fascinating but then asks me, "Which is bigger, a gene or a cell?" I think, Uh-oh, I really blew it. If I didn't make clear the basic biofact that while cells are certainly very small, each one is big enough to hold the entire complement of our 25,000 or so genes — as well as abundant bundles of tagalong genetic sequences, the function of which remains unknown — then what good am I? And when a copy editor, in the course of going over a story I've written about whale genetics, asks me to confirm the suggestions in my text that (a) whales are mammals and (b) mammals are animals, I think, Uh-oh, but this time in bold, twenty-six-point, panic-stricken type. Woe, woe, nobody knows anything about science. Woe, woe, nobody cares.

Am I sounding self-pitying, a sour-grapes-turned-defensive whine? Of course: a good offense begins with a nasal defensiveness. If I was going to write a book about the scientific basics, I had to believe that there was a need for such a book, and I do. If I believed there is a need for a primer, a guided whirligig through the scientific canon, then obviously I must believe there to be a large block of unprimed real estate in the world, vast prairies and deep arroyos of scientific ignorance and scientific illiteracy and technophobia and eyes glazing over and whales having their nursing privileges rescinded. In the civic imagination, science is still considered dull, geeky, hard, abstract, and, conveniently, peripheral, now, perhaps, more than ever. In a 2005 survey of 950 British students ages thirteen through sixteen, for example, 51 percent said they thought science classes were "boring," "confusing," or "difficult" — feelings that intensified with each year of high school. Only 7 percent thought that people working in science were "cool," and when asked to pick out the most famous scientist from a list of names that included Albert Einstein and Isaac Newton, many respondents instead chose Christopher Columbus.

Scientists are quick to claim mea culpas, to acknowledge that they bear some responsibility for the public allergy toward their profession. We've failed, they say. We've been terrible at communicating our work to the masses, and we're pathetic when it comes to educating our nation's youth. We've been too busy with our own work. We have to publish papers. We have to write grant proposals. We're punished by "the system," the implacable academic track that rewards scientists for focusing on research to the exclusion of everything else, including teaching or public outreach or writing popular books that get made into *Nova* specials. Besides, very few of us are as tele-elegant as Brian "String King" Greene, are we? All of which amounts to: guilty as charged. We haven't done our part to enlighten the laity.

A fair question to interject here is: Need we do anything at all? Does it matter if the great majority of people know little or nothing about science or the scientific mindset? If the average Joe or Sophie doesn't know the name of the closest star (the sun), or whether tomatoes have genes (they do), or why your hand can't go through a tabletop (because the electrons in each repel each other), what difference does it make? Let the specialists specialize. A heart surgeon knows how to repair an artery, a biologist knows how to run a gel, a jet pilot knows how to illuminate the FASTEN SEAT BELT sign at the exact moment you've decided to get up and go to the bathroom. Why can't the rest of us clip our coupons and calories in peace?

The arguments for greater scientific awareness and a more comfortable relationship with scientific reasoning are legion, and many have been flogged so often they're beginning to wheeze. A favorite thesis has it that people should know more about science because many of the vital issues of the day have a scientific component: think global warming, alternative energy, embryonic stem cell research, missile defense, the tragic limitations of the dry cleaning industry. Hence, a more scientifically sophisticated citizenry would be expected to cast comparatively wiser votes for Socratically wise politicians. They would demand that their elected representatives know the differences between a blastocyst, a fetus, and an orthodontist, and that one is a five-day-old, hollow ball of cells from which coveted stem cells can be extracted and theoretically inveigled to grow into the body tissue or organ of choice; the next is a developing prenate that has implanted in the mother's uterus; and the third is never covered by your company's dental plan.

Others propose that a scientifically astute public would be relatively shielded against superstitious, wishful thinking, flimflammery, and fraud. They would realize that the premise behind astrology was ludicrous, and that the doctor or midwife or taxi driver who helped deliver you exerted a greater gravitational pull on you at your moment of birth than did the sun, moon, or any of the planets. They would accept that the fortune in their cookie at the Chinese restaurant was written either by a computer or a new hire at the Wonton Food factory in Queens. They would calculate their odds of winning the lottery, see how ridiculously tiny they were, and decide to stop buying lottery tickets, at which point the education budgets of at least thirty of our fifty states would collapse. This last figure, alas, is not a joke, suggesting that if a pandemic of rational thinking should suddenly grip our nation, politicians might have to resort to dire measures to replace the income from state lotteries and state-owned slot machines, including — bwah-ha-ha! — raising taxes.

Lucy Jones, a seismologist at the California Institute of Technology, knows too well how resistant people can be to reason, and how readily they dive down a rabbit hole in search of axioms, conspiracy theories, the rabbit's fabled foot. A hearty, fiftyish woman with short, peach-colored hair and a rat-a-tat cadence, Jones serves as the United States Geological Survey's "scientist-in-charge" for all of Southern California, in which capacity she promotes the cause of earthquake preparedness. She has also been a designated USGS punching bag, officiating at media squalls and confronting public panic whenever the continental plate on which Southern California is perched gives a nasty shake. Like seismologists everywhere, she is trying to improve geologists' ability to predict major earthquakes, to spot the early warning signs in time to evacuate cities or otherwise take steps to protect people, their domiciles, that treasured set of highball glasses from the 1964 World's Fair. Jones has heard enough earthquake myths to shake a trident at: that fish in China can sense when a temblor is coming, for instance, or that earthquakes strike only early in the morning. "People tend to remember the early-morning earthquakes because those are the ones that woke them up and scared them the most," Jones said. "When you show them the data indicating that, in fact, an earthquake is as likely to happen at six P.M. as six A.M., they still insist there must be some truth to the story because their mothers and grandmothers and great-uncle Milton always said it was true. Or they will redefine 'early morning' to mean anything from midnight until lunchtime. And, by gosh, it's true: many earthquakes that occur, occur between twelve A.M. and twelve P.M. Uncle Milton was right!"

The public also believes that seismologists are much better at predicting earthquakes than they claim, but that they perversely keep their prognostications to themselves because they don't want to "stir a panic."

"I got a letter from a woman saying, 'I know you can't tell me when the next earthquake is going to be,'" Jones said, "'but will you tell me when your children go to visit out-of-town relatives?' She assumed I'd quietly use my insider's knowledge on behalf of my own family, while denying it to everybody else. People would rather believe the authorities were lying to them than to accept the uncertainty of the science." With a minimum of scientific training, Jones said, people would realize that the words "science" and "uncertainty" deserve linkage in a dictionary and that the only reason she would send her children to visit out-of-town relatives would be to visit out-of-town relatives.

Many scientists also argue that members of the laity should have a better understanding of science so they appreciate how important the scientific enterprise is to our nation's economic, cultural, medical, and military future. Our world is fast becoming a technical Amazonia, they say, a pitiless panhemispheric habitat in which being on a first-name basis with scientific and technical principles may soon prove essential to one's socioeconomic survival. "Soon after the Industrial Revolution, we in the West reached a point where reading was a fundamental process of human communication," Lucy Jones said. "If you couldn't read, you couldn't participate in ordinary human discourse, let alone get a decent job.

"We're going through another transformation in expectations right now," she continued, "where reasoning skills and a grasp of the scientific process are becoming things that everybody needs."

Scientists are hardly alone in their conviction that America's scientific eminence is one of our greatest sources of strength. Science and engineering have given us the integrated circuit, the Internet, protease inhibitors, statins, spray-on Pam (it works for squeaky hinges, too!), Velcro, Viagra, glow-in-the-dark slime, a childhood vaccine syllabus that has left slacker students with no better excuse for not coming to class than a "persistent Harry Potter headache," computer devices named after fruits or fruit parts, and advanced weapons systems named after stinging arthropods or Native American tribes.

Yet the future of our scientific eminence depends not so much on any cleverness in applied science as on a willingness to support basic research, the pi-in-the-sky investigations that may take decades to yield publishable results, marketable goodies, employable graduate students. Scientists and their boosters propose that if the public were more versed in the subtleties of science, it would gladly support generous annual increases in the federal science budget; long-term, open-ended research grants; and sufficient investment in infrastructure, especially better laboratory snack machines. They would recognize that the basic researchers of today help generate the prosperity of tomorrow, not to mention elucidating the mysteries of life and the universe, and that you can't put a price tag on genius and serendipity, except to say it's much bigger than Congress's science allotment for the current fiscal year.

Yes, let's cosset the scientists of today and let's home-grow the dreamers of tomorrow, the next generation of scientists. For by fostering a more science-friendly atmosphere, surely we would encourage more young people to pursue science careers, and keep us in fighting trim against the ambitious and far more populous upstarts India and China. We need more scientists! We need more engineers! Yet with each passing year, fewer and fewer American students opt to study science. As a National Science Board advisory panel warned Congress in 2004, "We have observed a troubling decline in the number of U.S. citizens who are training to become scientists and engineers," while the number of jobs requiring such training has soared. At this point, a third or more of the advanced science and engineering degrees earned each year in the United States are awarded to foreign students, as are more than half of the postdoctoral slots. And while there is nothing wrong with the international complexion that prevails in any scientific institution, foreign students often opt to take their expertise and credentials back to their grateful nation of origin. "These trends," the Science Board said, "threaten the economic welfare and security of our country."

Who can blame Americans for shunning science when, for all the supposed market demand, research jobs remain so poorly paid? After their decade or more of higher education, postdoctoral fellows can expect to earn maybe $40,000; and even later in their careers, scientists often remain stubbornly in the stratum of the five-figure salary. David Baltimore, a Nobel laureate and the former president of Caltech, who spent much of his early career at MIT, observed that the classic bakery for an upper-crust life, Phillips Academy prep school in Andover, Massachusetts, where his daughter was a student, has an excellent science program, one of the best. "But you never see Andover graduates at MIT," he said. "Academy alumni with quantitative skills go on to become stockbrokers. There are damned few patrician scientists."

Beyond better pay, science needs more cachet. Science advocates insist that if science were seen as more glamorous, racier, and more avant-garde than it is today, it might attract more participants, more brilliant young minds and nimble young fingers willing to click pipettes for twenty hours at a stretch. "Things were different while I was growing up," said Andy Feinberg, a geneticist at Johns Hopkins University. "It was the time of *Sputnik*, the race into space, and everybody was caught up in science. They thought it was important. They thought it was exciting. They thought it was cool. Somehow we must reinvigorate that spirit. The culture of discovery drives our country forward, and we can't afford to lose it."

These are all important, exciting, spirited arguments for promoting greater scientific awareness. I'd love to see more young Americans become scientists, especially the girl who serves as the vessel of my DNA and as a deduction on my tax return. I'd also be happy to see voters make smarter and more educated choices in Novembers to come than they have in the past.

And yet. As Steven Weinberg, a Nobel laureate and professor of physics at the University of Texas, points out, many issues of a supposedly scientific slant cannot be decided by science at all. "When it comes to something like the debate over an antiballistic missile defense system," he said, "I've been more bothered by the fact that our leaders seem to be the sort of people who don't read history rather than by the fact that they don't understand X-ray lasers." Can science really decide an issue like whether we should extract stem cells from a human blastocyst? All science can tell you about that blastocyst is, yep, it's human. It has human DNA in it. Science cannot tell you how much gravitas that blastocyst should be accorded. Science cannot settle the debate over the relative "right" of a blastocyst to its cellular integrity and uncertain future — deep freeze for possible implantation in a willing womb at some later date? or a swift bon voyage down the fertility clinic drainpipe? — versus the "right" of a patient with a harrowing condition like multiple sclerosis or Parkinson's disease to know that scientists have unfettered, federally financed access to stem cells and may someday spin that access into new therapies against the disease. This is a matter of conscience, politics, religious conviction, and, when all else fails, name-calling.

In sum, I'm not sure that knowing about science will turn you into a better citizen, or win you a more challenging job, or prevent the occasional loss of mental faculties culminating in the unfortunate purchase of a pair of white leather pants. I'm not a pragmatist, and I can't make practical arguments of the broccoli and flossing kind. If you're an adult nonscientist, even the most profound midlife crisis is unlikely to turn you into a practicing scientist; and unless you're a scientist, you don't *need* to know about science. You also don't need to go to museums or listen to Bach or read a single slyly honied Shakespeare sonnet. You don't need to visit a foreign country or hike a desert canyon or go out on a cloudless, moonless night and get drunk on star champagne. How many friends do you need?

In place of civic need, why not neural greed? Of *course* you should know about science, as much as you've got the synaptic space to fit. Science is not just one thing, one line of reasoning or a boxable body of scholarship, like, say, the history of the Ottoman Empire. Science is huge, a great ocean of human experience; it's the product and point of having the most deeply corrugated brain of any species this planet has spawned. If you never learn to swim, you'll surely regret it; and the sea is so big, it won't let you forget it.

Of course you should know about science, for the same reason Dr. Seuss counsels his readers to sing with a Ying or play Ring the Gack: These things are fun, and fun is good.

There's a reason why science museums are fun, and why kids like science. Science is fun. Not just gee-whizbang "watch me dip this rose into liquid nitrogen and then shatter it on the floor" fun, although it's that, too. It's fun the way rich ideas are fun, the way seeing beneath the skin of something is fun. Understanding how things work feels good. Look no further — there's your should.

"I was in college and in a debate with my father," said David Botstein, a geneticist at Princeton University. "He wanted me to be a doctor. I wanted to be a scientist. I had made it pretty clear to him that I wasn't going to medical school, and in fact I was already engaged in some really interesting research on DNA. One evening, a buddy of my father's, a general surgeon, cross-examined me about what it was I planned to do. How could anything be more interesting than human physiology and putting together broken bones? We were both having a little drink, and I explained to him what the structure of DNA meant, and its implications. This was back around 1960, when the field of molecular biology was just getting started. At the end of our conversation, my father's friend looks up, and says, 'You are the luckiest guy in the world. You are going to get paid to have fun.'"

Peter Galison, a professor of the history of physics at Harvard University, marvels cheekily at the thoroughness with which the public image of science has been drained of all joy. "We had to work really hard to accomplish this spectacular feat, because I've never met a little kid who didn't think science was really fun and really interesting," he said. "But after years of writing tedious textbooks with terrible graphics, and of presenting science as a code you can't crack, of divorcing science from ordinary human processes that use it daily, guess what: We did it. We persuaded a large number of people that what they once thought was fascinating, fun, the most natural thing in the world, is alien to their existence."

Granted, all the scientists I interviewed who attested to the fun of science are safely and amply granted, are flourishing in their fields and have personal cause to think the universe is a magical place. Yet I know plenty of very successful writers who think of themselves, not as the luckiest hey-you-guys in the world, but as cursed, as miserable, as being in their trade because they have no choice, no other marketable skills. "A writer is somebody for whom writing is more difficult than it is for other people," the novelist and essayist Thomas Mann complained. "When I come home for lunch after writing all morning, my wife says I look like I just came home from a funeral," said Carl Hiaasen — and he writes comic novels. David Salle, the artist, moaned to Janet Malcolm of *The New Yorker* about the miseries of painting. "I find it extremely difficult. I feel like I'm beating my head against a brick wall," he said. "I feel that everyone else has figured out a way to do it that allows him an effortless, charmed ride through life, while I have to stay in this horrible pit of a room, suffering." For their part, scientists are extremely bright and driven and — don't let their shorts and T-shirts fool you — carnivorously competitive; yet through it all they gush about the good fortune and great fun of being scientists, and they're not selfish and they're willing to share their glee.

"So, yes, we did it, we pushed the boulder to the top of the hill, and we made people think science is boring," Galison continued. But there's something to be said for a boulder in that position: it holds a lot of potential energy, and it's practically begging to be dislodged. A few well-placed shoves, a joining of shoulders for a hearty oomph, and the boulder may well be released from its unnatural bondage, to tumble earthward with a Newtonian roar.

This book is my small attempt to lend a deltoid to the cause, of nudging the boulder and unleashing the kinetic beauty of science to wow as it will.

Maybe you're one of those people who hasn't clicked with science since that dreadful year of high school when you flunked physics because you showed up for the final exam an hour late, in your pajamas, and carrying an insect collection. Or maybe you fulfilled your college science requirements by taking courses like the Evolutionary Psychology of Internet Dating, and you regret that you still can't tell the difference between a proton, a photon, and a moron. Or maybe you're just curiouser and curiouser and you don't know where to start. You think that the beginning might be a reasonable place, but whose beginning? Not the kiddie beginning, not the contemptuous or embarrassing or didactic digit-wagging beginning, but the beginning as an adult. The beginning as a relationship between equals, you and science. And before you raise your hands defensively, and cry, Whoa, that's not a fair competition, me versus science, let me say, It's not you *against* science, but you *with* science, you the taxpayer who supports science whether you realize it or not, you the person who does science more often than you'd suspect. Every time you try to isolate a problem with the vacuum cleaner, for example — machine heats up; machine stops running; holy hairball, when was the last time you changed the bag in this thing, anyway? Or when you know that if you don't stir the hollandaise sauce constantly at a hot but not boiling temperature you'll end up with a mass too lumpy to pour over your asparagus. You do science, you support science, you're baking the cake, you may as well lick the spoon.

This beginning is the beginning as scientists see it, or at least as they've agreed to see it because some reporter has shown up at their office door, plunked herself down in a chair, and asked them to consider a few very basic questions. Scientists have long whinnied about rampant scientific illiteracy and the rareness of critical thinking and the need for a more scientifically sophisticated citizenry. Fair enough. But what would it take to rid people of this dread condition, this pox populi ignoramus, and replace it with the healthy glow of erudition? What would a nonscientist need to know about science to qualify as scientifically seasoned? If you, Dr. Know, had to name a half-dozen things that you wish everybody understood about your field, the six big, bold, canonical concepts that even today still bowl you over with their beauty, what would they be? Or if you're the type of professor who still on occasion teaches undergraduate courses for those soft-shelled specimens known as "nonmajors," what are the essential ideas that you hope your students distill from the introductory class, and even retain for more than a few femtoseconds after finals? What does it mean to think scientifically? What would it take for a nonscientist to impress you at a cocktail party, to awaken in you the sensation that hmm, this person is not a buffoon?

When confronted with the query "What do you wish people knew about science?" many scientists felt compelled to talk about the urgent need to improve science education in primary and secondary school, which is a noble and necessary goal and worth urging at all relevant opportunities, but few adults have the luxury of a K-through-12 encore. To the well-intentioned curriculum revisionists, I gave my emphatic agreement, then pleaded that they take pity on the post-pedagogued. Surely not even the most feebly educated adult is beyond hope? Let's focus on them: What should nonspecialist nonchildren know about science, and how should they know it, and what is this thing called fun?

Realizing that the term "science" is a bit of a bounder, which can be induced via modifiers like "social" or "soft" to embrace anthropology, sociology, psychology, economics, politics, geography, or feng shui, I decided to focus on those sciences generally awarded the preamble "hard." These are the physical and life sciences, which in their broadest categories include physics, chemistry, biology, geology, and astronomy. These are the subjects that people tend to find the most daunting and abstruse, and that have the worst customer service desks. At the same time, they are the fields in which the greatest progress has been made, where the discoveries of the last century have been the grandest and most buoyant, and where a shopworn term like "revolutionary" still rightly applies. Scientists have probed the Joycean chambers of the atom, read the memoirs of the cosmos virtually back to the moment of crowning, detangled the snarls of our DNA, and mapped the twitchy globe of Silly Putty we call our castle and our home. These are the fairy tales of science, tales, as one scientist put it, "that happen to be true." They are hard the way diamonds and rubies are hard: they're built to last, and they sure look swell in the light.

In the course of my research, I interviewed and gathered insights from hundreds of scientists, often in person, sometimes by phone and e-mail, at many of the nation's premier universities and institutions. I spoke with Nobel laureates, members of the National Academy of Sciences, university presidents, institute directors, MacArthur geniuses. I also sought out researchers who were known as brilliant teachers, who had won their university's version of the "most adored professor of the year" award, or who were cited on student Web sites for being exceptionally clear, inspirational, entertaining, or, that old reliable, "awesome." Even the most difficult, desultory conversations, the ones that had me feeling like a Victorian dentist — all pliers and no nitrous — almost invariably yielded a gem or two. Scientists talked about the need to embrace the world as you find it, not as you wish it to be. They described their favorite molecules. They told jokes, like the one about physicist Werner Heisenberg, whose famed uncertainty principle says that you can know the position of an electron as it orbits the nuclear heart of an atom, or you can know its velocity, but that you can't know both at once. To wit: Heisenberg is scheduled to give a lecture at MIT, but he's running late and speeding through Cambridge in his rental car. A cop pulls him over, and says, "Do you have any idea how fast you were going?"

"No," Heisenberg replies brightly, "but I know where I am!"

"Now, you tell that at a cocktail party, and people will walk away from you," said Michael Rubner, a materials scientist at MIT. "Tell it in front of five hundred eighteen-year-olds at MIT, and they just roar."

I also pushed scientists to get beyond the knee-jerk tutorials, to explain, as much as was possible, what exactly they mean by some of the terms so often used as introductory definitions. You've likely heard, for example, the purportedly kindergarten description of the atom, that it is composed of three different classes of particles: protons and neutrons sitting sunlike at the center, electrons whizzing in orbits around them. You might also have heard that protons have a "positive charge," electrons a "negative charge," and neutrons "no charge." Well, that sounds breezy enough: a plus sign, a minus sign, and free with purchase. But what in the name of Mr. Rogers's last cardigan are we really talking about? What does it mean to say that a particle has "charge," and how does this subatomic "charge" of the light brigade relate to more familiar, real-world displays of electric "charge"? When your car breaks down in the middle of nowhere, for example, and you realize, on taking out your cell phone to call for help, that you forgot to re-"charge" the battery, and suddenly it's not a beautiful day in the neighborhood after all?

I also sought, as much as possible, to make the invisible visible, the distant neighborly, the ineffable affable. If a human cell were blown up to the size of something you could display on your coffee table, would you want to? What would it look like? You say that the average cell is a very busy place. Is that busy like Manhattan, or busy like Toronto?

It's not that I wanted to take dumbing-down to new heights. In peppering sources with the most pre-basic of questions and tapping away at the Plexiglas shield of "everybody knows" until I was about as welcome as a yellow jacket at a nudist colony, I had several truly honorable aims. For one thing, I wanted to understand the material myself, in the sort of visceral way that allows one to feel comfortable explaining it to somebody else. For another, I believe that first-pass presumptions and nonexplanatory explanations are a big reason why people shy away from science. If even the Shlemiel's Guide to the atom begins with a boilerplate trot through concepts that are pitched as elementary and self-evident but that don't, when you think about them, really mean anything, what hope is there for mastering the text in cartoon balloon number two?

Moreover, in choosing to ask many little questions about a few big items, I was adopting a philosophy that lately has won fans among science educators — that the best way to teach science to nonscientists is to go for depth over breadth.

After countless interviews and many months of labor, I began to experience the wonderful, terrible sensation of "déjà-knew": scientists were telling me the same things I'd heard before. Wonderful, because it meant I could be fairly confident I had a defensible corpus of scientific fundamentals that weren't entirely arbitrary or idiosyncratic. Terrible, because it meant the time for reporting was over, and the time had arrived for writing, the painful process, as the neuroscientist Susan Hockfield so pointedly put it, of transforming three-dimensional, parallel-processed experience into two-dimensional, linear narrative. "It's worse than squaring a circle," she said. "It's squaring a sphere." And to think I was brought to tears in an art class because I couldn't draw a straight line.

1

Thinking Scientifically

An Out-of-Body Experience

SCOTT STROBEL, A BIOCHEMIST at Yale University, is tall, tidy, and boyishly severe, his complexion a polished apple, his jaw ajut, his hair a sergeant's clipped command. He looks athletic. He keeps pictures of his three beaming children on his desk. I am not surprised to learn that he graduated summa cum laude from Brigham Young University. He might be good company at a family picnic, but on this fluorescent-enhanced midweek morning, as we sit around his office coffee table engaged in what he has deemed a form of constructive entertainment, Strobel is about as much fun as an oncologist.

Strobel has taken out his personal kit of Mastermind, a game I had never seen before and knew nothing about. He often plays the game with the graduate students and postdoctoral fellows in his lab. They love it. So, I later discovered, do my husband and daughter. Now Strobel is teaching me to play Mastermind, but of the many words competing for the tip of my tongue, "love" is not one of them.

In Mastermind, he explains, you try to divine your opponent's hidden sequence of four colored pegs by shuffling your own colored pegs among peg holes. If you guess a correct color in the correct position, your opponent inserts a black peg on his side of the board; a correct color in an incorrect position gets you a white peg; and the wrong color for any position earns you no peg at all. Your goal is to end up with four black pegs on your adversary's end in as few rounds as possible.

"Got it?" he says, pushing the board in my direction.

"I never really liked games," I plead. "Don't you have any nice slide presentations instead?"

"I have a point to make with this," he says. "Go ahead."

Without a tornado or the sudden onset of pneumococcal pneumonia to deliver me, I sigh and arrange my pegs in a pleasant police lineup of blue, red, yellow, green. Strobel responds in a pattern of blacks, whites, and blanks. I lunge with a red piece, he parries by plucking off a white peg. Green here? Sorry, dear. I'm trying my best, but I have a wooden ear for the game, and I make bad choices and no progress. I fight back tears, which fecklessly leap to freedom as sweat. I curse Strobel and all scientists who ever lived, especially the inventor of the pegboard.

Finally, Strobel takes pity on me. "Well, I think you get the idea," he says. He sweeps the malignant little pins back into their box, and I lapse into limp remission.

Mastermind, he declares, is "a microcosm for how science works." By insisting I play the game, he was trying to impress on me an essential truth about science. And while the dramady at Strobel's gaming table was not my favorite hour, in its intensity and memorability it reflects the strength with which scientists, whatever their specialty, agree with this truth.

Science is not a body of facts. Science is a state of mind. It is a way of viewing the world, of facing reality square on but taking nothing on its face. It is about attacking a problem with the most manicured of claws and tearing it down into sensible, edible pieces.

Even more than the testimonials to the fun of science, I heard the earnest affidavit that science is not a body of facts, it is a way of thinking. I heard these lines so often they began to take on a bodily existence of their own.

"Many teachers who don't have a deep appreciation of science present it as a set of facts," said David Stevenson, a planetary scientist at Caltech. "What's often missing is the idea of critical thinking, how you assess which ideas are reasonable and which are not."

"When I look back on the science I had in high school, I remember it being taught as a body of facts and laws you had to memorize," said Neil Shubin, a paleontologist at the University of Chicago. "The Krebs cycle, Linnaean classifications. Not only does this approach whip the joy of doing science right out of most people, but it gives everyone a distorted view of what science is. Science is not a rigid body of facts. It is a dynamic process of discovery. It is as alive as life itself."

"I couldn't care less whether people memorize the periodic table or not," said David Baltimore, the former president of Caltech. "I understand they're more concerned with problems that are meaningful in their own lives. I just wish they would approach those problems in a more rational way."

When science is offered as a body of facts, science becomes a glassy-eyed glossary. You skim through a textbook or an educational Web site, and words in boldface leap out at you. You're tempted to ignore everything but the highlighted hand wavers. You think, if I learn these terms, maybe I won't flunk chemistry. Yet if you follow such a strategy, chances are excellent that you will flunk chemistry in the ways that matter — not on the report card in your backpack, but on the ratings card in your brain.

The conjuring of science as a smarty-pants set of unerring facts that might be buzzed up on a *Jeopardy!* afternoon also suits the opponents of science, like the antievolutionists who seize on every disputed fossil to question the entire Darwinian enterprise. "Creationists first try to paint science as a body of facts and certainties, and then they attack this or that 'certainty' for not being so certain after all," said Shubin. "They cry, 'Aha! You can't make up your mind. You can't be trusted. Why should we believe you about anything?' Yet they are the ones who constructed the straw man of scientific infallibility in the first place."

"Science is not a collection of rigid dogmas, and what we call scientific truth is constantly being revised, challenged, and refined," said Michael Duff, a theoretical physicist at the University of Michigan. "It's irritating to hear people who hold fundamentalist views accuse scientists of being the inflexible, rigid ones, when usually it's the other way around. As a scientist, you know that any new discovery you're lucky enough to uncover will raise more questions than you started with, and that you must always question what you thought was correct and remind yourself how little you know. Science is a very humble and humbling activity.

"Which doesn't mean," Duff added hastily, "that there aren't arrogant scientists around."

Back at Yale University, Strobel further explains the message of Mastermind. If science is not a static body of facts, what is it? What does it mean to think scientifically, to take a scientific whack at a problem? The world is big. The world is messy. The world is a teenager's bedroom: Everything's in there. Now how do you get it to the kitchen sink? How can you possibly begin to make sense of it? One furred fork, one accidental petri dish, one peg hole at a time.

"If you're trying to pose a question in a way that gets you data you can interpret, you want to isolate a variable," Strobel says. "In science we take great pains to design experiments that ask only one question at a time. You isolate a single variable, and then you see what happens when you change that variable alone, while doing your best to keep everything else in the experiment unchanged." In Mastermind, you change a single peg and watch the impact of that deviation on your "experiment." In science, if you'd like to know, for example, whether a chemical reaction depends on the presence of oxygen, you would stage the experiment twice, first with oxygen, then without. Everything else you'd keep the same to the closest approximation possible — same heat, same light, same timing, same type of container; and, just to be safe, same white socks and Tevas.

You don't need to work at a laboratory bench to follow a scientific game plan. People behave scientifically all the time, although they may not realize it. "If someone is trying to fix a DVD player, they do experiments, they do controls," said Paul Sternberg, a developmental biologist at Caltech. "Step one is observation: What does the picture look like? What are the possible things that could be wrong here? Is it really the player, or could it be the television set? You come up with a hypothesis, then you start testing it. You borrow your neighbor's DVD player, you hook it up, you see your TV set is fine. So you check your DVD's input, output, a couple of wires. You may be able to track down the problem even without really understanding how a DVD player works.

"Or maybe you're trying to troubleshoot your pet," Sternberg said. "Why does the fish look funny? Why is my dog upset? I'll feed the hamster less or I'll feed it more, or maybe it doesn't like the noise, so I'll move it away from the stereo system. Should I take Job A or Job B? Well, let me see how long the drive would be from the office to my daughter's school during rush hour; that could be the killer factor in making a decision. These are all examples of forming hypotheses, doing experiments, coming up with controls. Some people learn these things at an early age. I had to get a Ph.D. to figure them out."

A number of scientists proposed that people may have been more comfortable with the nuts and bolts of science back when they were comfortable with nuts and bolts. "It was easier to introduce students and the lay public to science when people fixed their own cars or had their hands in machinery of various kinds," said David Botstein of Princeton. "In the immediate period after World War II, everybody who'd been through basic training knew how a differential gear worked because they had taken one apart."

Farmers, too, were natural scientists. They understood the nuances of seasons, climate, plant growth, the do-si-do between parasite and host. The scientific curiosity that entitled our nation's Founding Fathers to membership in Club Renaissance, Anyone? had agrarian roots.

Thomas Jefferson experimented with squashes and broccoli imported from Italy, figs from France, peppers from Mexico, beans collected by Lewis and Clark, as he systematically sought to select the "best" species of fruits and vegetables the world had to offer and "to reject all others from the garden." George Washington designed new methods of fertilizing and rotating crops and invented the sixteen-sided treading barn, in which horses would gallop over freshly harvested wheat and efficiently shake the grain from the stalks.

"The average adult American today knows less about biology than the average ten-year-old living in the Amazon, or than the average American of two hundred years ago," said Andrew Knoll, a professor of natural history at Harvard's Earth and Planetary Sciences Department. "Through the fruits of science, ironically enough, we've managed to insulate people from the need to know about science and nature." Yet still, people troubleshoot their pets, their kids, and, in moments of utter recklessness, their computers, and they apply scientific reasoning in many settings without realizing it, for the simple reason that the method works so well.

Much of the reason for its success is founded on another fundamental of the scientific bent. Scientists accept, quite staunchly, that there is a reality capable of being understood, and understood in ways that can be shared with and agreed upon by others. We can call this "objective" reality if we like, as opposed to subjective reality, or opinion, or "whimsical set of predilections." The contrast is deceptive, however, for it implies that the two are discrete entities with remarkably little in common. Objective reality is out there, other, impersonal, and "not me," while subjective reality is private, intimate, inimitable, and life as it is truly lived. Objective reality is cold and abstract; subjective reality is warm and Rockwell. Science is effective because it bypasses such binaries in favor of what might be called empirical universalism, the rigorously outfitted and enormously fruitful premise that the objective reality of the universe comprises the subjective reality of every one of us. We are of the universe, and by studying the universe we ultimately turn the mirror on ourselves. "Science is not describing a universe out there, and we're separate entities," said Brian Greene. "We're part of that universe, we're made of the same stuff as that universe, of ingredients that behave according to the same laws as they do elsewhere in the universe."

A molecule of water beaded on a forehead at Yale University would be indistinguishable from a molecule of water skating through space aboard Comet Kohoutek. Ashes to ashes, stardust to our dust. As I'll describe later in detail, the elements of our bodies, and of the earth, and of a painted Grandma's holiday apron, were all forged in the bellies of long-dead suns.

To say that there is an objective reality, and that it exists and can be understood, is one of those plain-truth poems of science that is nearly bottomless in its beauty. It is easy to forget that there is an objective, concrete universe, an outerverse measured in light years, a microverse trading in angstroms, the currency of atoms; we've succeeded so well in shaping daily reality to reflect the very narrow parameters and needs of *Homo sapiens.* We the subjects become we the objects, and we forget that the moon shows up each night for the graveyard shift, and we often haven't a clue as to where we might find it in the sky. We are made of stardust; why not take a few moments to look up at the family album? "Most of the time, when people walk outside at night and see the stars, it's a big, pretty background, and it's not quite real," said the Caltech planetary scientist Michael Brown. "It doesn't occur to them that the pattern they see in the sky repeats itself once a year, or to appreciate why that's true."

Star light, star bright, Brown wishes you'd try this trick at night: Pay attention to the moon. Go outside a few evenings in any given month, and see what time the moon rises, and what phase it's in, and when it sets, and then see if you can explain why. "Just doing this makes you realize that the sun and moon are both out there," he said, "and that the sun is actually shining on the moon, and the moon is going around the Earth, and that it's not all a Hollywood special effect." Brown knows first-eye how powerful such simple observations can be. It was the summer after he'd graduated from college, and he was biking across Europe and sleeping outside each night. In accordance with his status as young, footloose, and overseas, he wore no wristwatch, so he sought to keep time by the phases of the moon. "I realized that I had never noticed before that the full moon rises when the sun sets," he said. "I thought, Hey, you know, this makes sense. I suppose I should have been embarrassed not to have noticed it before, but I wasn't. Instead, it was just an amazing feeling. The whole physical world is really out there, and things are really happening. It's so easy to isolate yourself from most of the world, to say nothing of the rest of the universe."

The last spring of my father's life, before he died unexpectedly of a fast-growing tumor, he told me that it was the first time he had stopped, during his walks through Central Park in New York, and paid attention to the details of the plants in bloom: the bulging out of a bud from a Lenten rose, the uncurling of a buttery magnolia blossom, the sprays of narcissus, Siberian bugloss, and bleeding heart. I was so impressed by this that, ever since, I have tried to do likewise, attending anew to the world in rebirth. Each spring I ask a specific question about what I'm seeing and so feel as though I am lighting a candle in his memory, a small focused flame against the void of self-absorption, the blindness of I.

Another fail-safe way to change the way you see the world is to invest in a microscope. Not one of those toy microscopes sold in most Science 'n' Discovery chain stores, which, as Tom Eisner, a professor of chemical ecology at Cornell, has observed, are unwrapped on Christmas morning and in the closet before Boxing Day. Not the microscopes that magnify specimens up to hundreds of times and make everything look like a satellite image of an Iowa cornfield. Rather, you should buy a dissecting microscope, also known as a stereo microscope. Admittedly, such microscopes are not cheap, running a couple of hundred dollars or so. Yet this is a modest price to pay for revelation, revolution, and — let's push this envelope out of the box while we're at it — personal salvation. Like Professor Brown, I speak from experience. I was accustomed to looking through high-powered microscopes in laboratories and seeing immune cells and cancer cells and frogs' eggs and kidney tissue from fetal mice. But it wasn't until my daughter received a dissecting microscope as a gift, and we began using it to examine the decidua of everyday life, that I began yodeling my hallelujahs. A feather from a blue jay, a fiddlehead fern, a scraping from a branch that turned out to be the tightly honeycombed housing for a stinkbug's eggs. How much heft and depth, shadow and thistle, leap out at you when the small is given scope to strut. At a mere 40× magnification, salt grains look like scattered glass pillows, a baby beetle becomes a Fabergé egg, and, as much as I hate mosquitoes, a mosquito under the microscope is pure Giacometti: *Thin Man Takes Wing, with Violin.*

Yes, the world is out there, over your head and under your nose, and it is real and it is knowable. To understand something about why a thing is as it is in no way detracts from its beauty and grandeur, nor does it reduce the observed to "just a bunch of" — chemicals, molecules, equations, specimens for a microscope. Scientists get annoyed at the hackneyed notion that their pursuit of knowledge diminishes the mystery or art or "holiness" of life. Let's say you look at a red rose, said Brian Greene, and you understand a bit about the physics behind its lovely blood blush. You know that red is a certain wavelength of light, and that light is made of little particles called photons. You understand that photons representing all colors of the rainbow stream from the sun and strike the surface of the rose, but that, as a result of the molecular composition of pigments in the rose, it's the red photons that bounce off its petals and up to your eyes, and so you see red.

"I like that picture," said Greene. "I like the extra story line, which comes, by the way, from Richard Feynman. But I still have the same strong emotional response to a rose as anybody else. It's not as though you become an automaton, dissecting things to death." To the contrary. A rose is a rose is a rose; but the examined rose is a sonnet.

That the universe can be explored and incrementally understood without losing its "magic" does not imply a corollary: that maybe "magic" is true after all, is hidden under accretions of apparent order, and that one of these days reality will kick off on a bucking broomstick toward Hogwarts on the hill. The universe still brims with mysteries, of course, but, in their conviction that the universe is knowable, scientists doubt that these question marks, once they have been understood well enough to become commas, will prove to be regions of arbitrary lawlessness or paranormality. "We have a pretty good idea of what kind of world this is, and it is not as mysterious, in the conventional sense of the word, as some people might wish," said Steven Weinberg. "It's not a world in which human destiny is linked to the positions of planets, or where people can be cured by crystals or bend spoons with their thoughts. Sometimes the police will call in a psychic to help solve a crime, and you'll hear a discussion on television for or against. But this isn't really an open question."

For example, one of the great conundrums in astronomy is the nature of something called dark energy, a kind of antigravitational force that appears to be pushing the accelerator pedal of the universe. The universe, as we'll discuss later, was born in the celebrated Big Bang about 13.7 billion years ago and has been expanding ever since; that much is clear and nearly incontrovertible. Yet until quite recently scientists thought that the rate of expansion was slowing down. You know how it is: a youthful burst of levity, and then the years start tugging on the back of your shorts. So, too, it was believed, for the universe: the gravitational pull of all its mass was supposed to be slowing down its rate of expansion. Instead, researchers have seen the opposite. The expansion is speeding up. Galaxies are flying away from one another at an ever increasing pace. Our universe has found a second wind. What is the meaning of this shadowy force, this type A provocateur, this energy so studiously seditious it hides behind dark glasses? Does its existence call into question the entire edifice of astrophysics, of what we've learned about the universe to date? To quote that most cerebral of comics, Steve Martin: "Nah!" Scientists are dazzled by dark energy. They are impressed by its size and strength. They want very, very much to understand it. Nobody I spoke with, however, felt threatened by it. They have some ideas about what dark energy may be. They're open to other, better suggestions. They're just not about to consult a psychic for help in finding the body.

After all, history is replete with "unfathomable" mysteries that have been fathomed into the archives. The physicist Robert Jaffe of MIT cited the case of what might be called spire and brimstone. The cathedrals and churches of Christendom traditionally were built on the highest promontory in town and outfitted with the loftiest steeples parishioners could afford, the better to reach toward heaven and vamp for the neighbors. Unfortunately, those tall, wooden towers attracted more than envy: churches were regularly struck by lightning and burned to varying degrees of a crisp. "Every time this happened, there would be a wrenching dialogue about sin and the vengeance of God," said Jaffe, "and what the parish had done to bring the wrath of the Lord upon them." Then, in the eighteenth century, Benjamin Franklin determined that lightning was an electric rather than an ecclesiastic phenomenon. He recommended that conducting rods be installed on all spires and rooftops, and the debates over the semiotics of lightning bolts vanished. Nowadays, a fire in a church is less likely to be considered an act of God than of a tippling priest who neglected to blow out the candles.

Scientists may believe that much, if not all, of the universe will prove comprehensible, yet interestingly, this comprehensibility continues to astound them. Immanuel Kant observed that "the most astonishing thing about the universe is that it can be understood." This was hardly a clause in a prenuptial agreement. As the Princeton astrophysicist John Bahcall put it in an interview shortly before he died, we crawled out of the ocean, we are confined to a tiny landmass circling a midsize, middle-aged, pale-faced sun located in one arm of just another pinwheel galaxy among millions of star-spangled galaxies; yet we have come to comprehend the universe on the largest scales and longest time frames, from the subatomic out to the edge of the cosmos. "It's remarkable, it's extraordinary, and it didn't have to be that way," Bahcall said.

In other words, we can count our lucky stars that the stars can be counted. "You can imagine a universe that's complicated no matter how you look at it or try to break it down," said Brian Greene. "But we don't live in that kind of universe, and I for one am grateful." The world may seem confusing, chaotic, unspeakably rude, yet underlying it all is a certain amount of order. "The wonder of science is that a few very simple ideas can yield incredibly rich phenomena," said Greene. "It's astounding that a few symbols on a blackboard underlie so much of what we experience." Ah, yes, "a few symbols on a blackboard," the smudged garden of glyphs that covered Greene's blackboard, and the green boards and the black-markered white boards of every physicist I visited. Physicists don't just scribble equations when they're posing for cartoonists. They scribble to one another, too. They talk the talk, they chalk the chalk, and they, like us, marvel at how often their abstract computations fit the fleshiness of life. The physicist Eugene Wigner talked of "the unreasonable effectiveness of mathematics" — in delineating the present, disinterring the past, and baking a trustier fortune cookie. With the aid of mathematics, scientists can calculate solar eclipses thousands of years in advance, for example, or gauge when to launch a space probe so that it will rendezvous with Neptune, or predict the life span and death throes of a distant star. Mathematics has proved to be such a potent means for dissecting reality that many scientists see it as not merely a human invention, like a microscope or a computer, but a reflection of traits inherent to the cosmos, a glimpse into its underlying architecture and operating system. By this view, you needn't be the hominid descendant of a lungfish or the intellectual descendant of the Greek mathematician Euclid to realize that the structure of space-time has a distinct saddleback geometry to it, which we earthlings label non-Euclidean. "When somebody says they were the first person to discover quantum mechanics or relativity or the like, I always think to myself, it's probably been discovered millions of times before, by other civilizations elsewhere in this galaxy or in other galaxies," said the theoretical physicist John Schwarz of Caltech.

For all the power of math in making sense of reality, though, math should not be thought of as something inviolate, matchless, even sacred. A mathematical description of a phenomenon is not a "truer" description than an equivalent, nonmathematical explanation would be, any more than the word "table" is a truer rendering of "a piece of furniture having a smooth, flat top on legs" than are the words "mesa," "tavolo," or "lijst." Math is *a* language, not *the* language, and its symbols can be explained in other idioms, including that lovely English dialect called Plain. For all but a tiny clique of researchers known as pure mathematicians, who have scant interest in connecting the dots between theorem and you-are-here, math is a means to an end, and the end must do more than make the pi higher. It must deliver reality back to us, this time with chapter headings, annotations and footnotes, and wise verbs strong enough to bear the weight of the inevitable sentence endpoint, the question mark. I get irritated with scientists who complain about the reluctance of popular science writers to include a sprinkling of math in their narrative, and who insist that the story told is therefore incomplete and even slightly misleading, as though the point of the math was the math was the math. "In principle, every equation can be expressed in English as a sentence," said Brian Greene. Admittedly, such transpositions often would be clumsy sentences, and you wouldn't want to curl up with a book of them, but the moral is clear: even if you remain numb to numbers, you can still understand what they have to tell us about the universe. You can become scientifically quite sophisticated without mastering much if any math. "I have never felt that science was quite so dependent on mathematics as some scientists do," said Kip Hodges, director of the School of Earth and Space Exploration at Arizona State University. "Mathematics is a way of describing nature but not necessarily of understanding it."

Yes, our children should be taught much more math and in far greater depth than they currently are in the average American classroom. Absolutely. But we must face the sad truth that children can take it, and adults cannot. As a consequence of brain biology, children are brilliant at learning new languages of all sorts. Their neurons are practically liquid, pouring across local loci and making new friends and synapses with hardly a grunt of effort. As we age, however, the cells settle into place, maybe invest in a sofa and china cabinet, and the entire neuronal matrix, slowly but unmistakably, starts to harden. By our late twenties or early thirties, the mind is made up: it has taken a stand on life, it knows from whence it speaks, and that commitment is reflected in its structure. Of course we can learn new things, up until the day we learn how to die; but chances are excellent that most adult learning takes place through the prism of preexisting skills. So if math is all Greek to you, take comfort in the following: (a) Why shouldn't it be? Many of the symbols used in math are letters from the Greek alphabet; and (b) it's Greek to a surprising number of scientists, too. As it happens, many biologists, chemists, geologists, and astronomers are relatively poor mathematicians. Bonnie Bassler of Princeton, considered one of the brightest young stars in the field of bacterial ecology, confessed to me that she is "terrible at math" and always has been. "I can balance my checkbook if I have a calculator," she said. "I can do fractions. But that's it. Somehow it didn't matter, and I ended up here."

Even physicists, for whom math is indispensable, have their limits. Steven Weinberg may have won a Nobel Prize for helping to develop the mathematics that merged two of nature's four fundamental forces, electromagnetism and the weak force, into a single theoretical bundle called the electroweak force — and this is not something you could do by reviewing your old high school algebra notes — yet he said he recently switched from particle physics to cosmology because the math in particle physics was getting beyond him.

Yet while a mastery of math is not essential to appreciating and even practicing science, you can't avoid, while milling through the fairground of Science Mind, bumping into a few cousins from math's extended family. One is quantitative thinking, to which the next chapter is devoted: becoming comfortable with concepts of probability and randomness, and learning a few tricks about how to break a problem into tractable pieces and to whip up a back-of-a-wet-cocktail-napkin estimate of some seemingly incalculable figure, like, how many school buses are in your county, or how many people would have to hold hands to form a human chain around the globe and how many of them will be bobbing in open ocean and had better bring a life jacket, shark repellent, and a copy of their dental records just in case? True, you can likely find the answers to these and other fun FAQs on the Internet, yet the habit of thinking in stepwise, quantitative fashion, and facing a problem head-on rather than running off screaming to Google, is worth cultivating. Second only to their desire that science be seen as a dynamic and creative enterprise rather than a calcified set of facts and laws, scientists wish that people would learn enough about statistics — odds, averages, sample sizes, and data sets — to scoff with authority at crooked ones. Through sound quantitative reasoning, they reason, people might resist the lure of the anecdote and the personal testimonial, the deceptive N, or sample size, of "me, my friends, the doorman, and the barista at Caribou." With a better appreciation for the qualities of quantities, people might be able to set aside, if only temporarily, the stubbornness of a human brain that evolved to focus on the quirks and peccadilloes of a small, homogeneous tribe, rather than on the daunting population densities and polycultural vortices that characterize life in contemporary Gotham City. There is a little principle called the law of large numbers, which among other things means that if the group you're considering is very big, nearly anything is possible. Events that would be rare on a limited scale become not merely common, but expected. One favorite example among the numerati is that of repeat lottery winners, people who have won big prizes two or more times and who invariably provoke clucks of awe, envy, what-are-the-odds. "The really amazing thing would be if nobody won twice," said Jonathan Koehler, a professor of economics at the University of Texas.

By thinking small in a large land, we get a skewed sense of what's meaningful and what's happenstance. "People are overly impressed by coincidences, and they get fooled by them," said John Allen Paulos, a mathematician at Temple University and the author of *Innumeracy* and many other books. Paulos has toyed with the idea of playing the Barnum card to make a point while making a profit. He could start a newsletter of random predictions about the stock market and mail it to two large sets of readers. One group would receive a newsletter predicting that the market would rise in the next three months; another would be told that the market would go bearish. Three months later, he'd see how the market had fared, and direct his next newsletter solely to the recipients of his correct first guess, again separating them into two camps. Half would be flagged to expect a bull market, and half would be warned of an imminent downturn. By the third newsletter, he could boast to a winnowed but still substantial pool of readers, Hey, I've successfully predicted the stock market for two cycles running, and then ask, Care to invest $10 to receive my next divination? (Keep Paulos's scheme in mind should you receive any suspicious solicitations from Temple University.)

Another aspect of quantitative reasoning that characterizes the scientific mindset is this: there must be some quantity to it, some substance, some evidence. Science demands evidence: Does this sound, well, self-evident? Maybe so, but it's a lesson that can be awfully hard to swallow, and must be taken again and again, our daily ABCs and periodic Mendeleevs, folic acid for the backbone, iron in homage to the core of the earth. It's hard to swallow because we love opinions. The most thoroughly read pages in a newspaper are the opinion pages — the editorials, the columns and commentaries, the bellicose lettres from readers living somewhere in the state of Greater Umbrage. Opinions are to have and to hold, in sickness and in health, over breakfast or by blog. Opinions feel good. You're entitled to yours; I'll indulge mine. "In politics, you can say, I like George Bush, or I don't like George Bush, or I do or don't like Howard Dean or John Kerry or Mr. Magoo," said Andrew Knoll of Harvard. "You don't need a principled reason for that political opinion. You don't need evidence that someone else can replicate to justify your opinion. You don't need to think of alternative explanations that would render your opinion invalid. You can go into the voting booth, and say, I prefer this or that politician, and cast your vote accordingly. You don't need excuses for the foods you like, either. If you're ordering dinner at a restaurant, you can ask that your steak be cooked rare or medium or well-done, and the waiter isn't likely to stop and demand that you present evidence to back up your taste, at least not if he wants his tip.

"Unfortunately, people often regard science the same way, as a matter of opinion," Knoll continued. "I do or don't like George Bush, I do or don't believe in evolution. It doesn't matter why I don't believe in evolution, it doesn't matter what the evidence is, I just don't believe in it." You, the evolutionist, "believe" in evolution; I, the creationist, do not. You have your opinion, I have mine, and it takes all kinds of nuts and dips to make a party, right?

At which point most evolutionists are likely to get very impatient and form opinions of their interlocutor that they may or may not choose to express. Scientists can be quite hard on one another, too. They sneer, they dismiss, they scrawl comments on one another's submitted reports like "I feel sorry for whoever funded this so-called research" or "I wouldn't publish this at the bottom of a birdcage." Yet for all the crude inanity of its more extreme sputterings, the attack-dog stance is part of science's strength. The big difference between science and many other aspects of life is, to quote George W. Bush's response to a disgruntled citizen at a July Fourth picnic, "Who cares what you think?" Your opinion doesn't count. Your fond hopes and fantasies of Paradigms Found don't count. What counts is the quality and the quantity of the evidence.

"How you want it to be doesn't make any difference," said the biologist Elliot Meyerowitz of Caltech. "In fact, if things are turning out the way you want them to, you should think harder about how you're doing your experiments, to make sure you're not introducing some bias." As members of the human race, scientists are born to be biased, particularly in favor of their personal biases. After all, we're stuck in our skulls for the whole four-score sentence of sentience. We can't brainhop or mindswap; we merely window-shop. I think, therefore I am right. Yet while self-delusion has been shown to be an extremely useful tool in many situations — particularly when trying to persuade a potential employer or love interest of your extraordinary worth — it is, in the words of the MIT molecular biologist Gerald Fink, "the enemy of science."

"Those of us who are not overly philosophical believe that there is a reality to nature but that it can be very hard to see it and understand it, given all our biases," Meyerowitz said. "The reason a scientist spends all those years in training, as an undergraduate, graduate student, and postdoc, is to learn to deal with personal biases." Good scientists spend a lot of time assuming they're up to no good. They are anti–Sixth Amendment, guilty until proven innocent, or penitents in search of redemption. "If you're doing your job," said the chemist Daniel Nocera of MIT, "you should be the one who disproves yourself most of the time." It doesn't matter what sort of story you tell yourself as you are doing your experiments, what hypothesis you formulated before you started clicking your pipette or infusing your fetal mice with fluorescent green marker from a jellyfish. Just make sure that the endpoints are pure of heart. "The results section of a scientific paper is where you show you're a good scientist. Here is where you say, I did the experiment properly, and collected the data properly and the data are right," said Nocera. "In the discussion section, where you talk about the implications of the work, you can sound smart or stupid, and tell an interesting story or not. I warn my students, you may sometimes be stupid and you may sometimes be smart, but you must always be good. When I read the results section of your paper, everything in there has got to be right." Darcy Kelley, a neuroscientist at Columbia, sounds a similar warning knell to her students: "Your data should be true even if your story is wrong."

How do scientists seek to purge their work of bias and bad data? Through frequent ablutions at the baptistry of the Control. As vital to the integrity of a scientific report as the finding being showcased are all the no-shows offered in comparison: We did operation A to variable B and got result Z; but when we subjected B to operations E, I, O, U, and even Y, B didn't budge. When researchers at Boston University wanted to show that the eggs of a red-eyed tree frog would hatch early expressly to avoid predation by an oncoming snake, allowing the preemie tadpoles to leap to safety in the water below, it wasn't enough to film the unripe eggs bursting open on the approach of an oviphagous serpent: after all, who's to say that the eggs were responding to a snake-specific threat rather than to an ambient disturbance? The scientists demonstrated the precision of the frog eggs' monitoring system by exposing them to a variety of recorded vibrations of equal amplitude from distinct sources — slithering snake, passing human footsteps, hammering rain. Only with a snake shake would the tadpoles make haste.

A lovable control is often blind: those who perform the experiment should be unaware of what's control and what's the real thing until all the results are in, at which stage the code can be broken. Sometimes devising the right controls is the hardest part of a study. When researchers sought to demonstrate the effectiveness of acupuncture to treat a variety of ailments — lower back pain, diabetes, depression — they yearned to be taken seriously. They were tired of their colleagues' twitchy-kneed rejection of all alternative healing practices, and they were really tired of the catty references to "quackupuncture." They wanted the fourteen-karat validation of a blinded study, in which one group of patients received acupuncture and one did not, and neither set would know who was the treated, who the placebo. But how to fool some of the people some of the time about a procedure as palpable as playing pincushion? The researchers' solution was dapper and to the point: one group of patients would be given needles inserted into officially designated acupuncture nodes, while the second group would have needles inserted into "sham" spots on the body that acupuncturists agreed should have no effect. When patients with lower back pain reported relief from bona fide needling but not from sham acupuncture, even the most skeptical Western doctors had to concede that the 5,000-year-old practice might have its limited uses.

"In my life as a scientist, the thing I worry about the most is, What are the right controls?" said Gerald Fink. "You send a paper off for publication, and you're stricken with doubt: Did I do it? Did I use the right controls?"

Another route to data security is . . . another route. Approach a problem from many angles and see if you always end up in Rome. One of my favorite examples of meticulous cartography is a report by Gene Robinson, a neuroethologist at the University of Illinois in Urbana-Champaign. Neuroethologists study the neurobiology of behavior, in Robinson's case of bee behavior. He's exploring how gene activity in the brain is linked to an individual's conduct, and he has decided that the best way to address these big, socially flammable questions is on the modest terrain of the bee brain, which would fit snugly into the belly of this *b*. His question: How does a bee know what to be and not to be? How does a worker bee know that she's meant to spend the first half of her six-week life performing hive-bound duties like tending to the eggs, cleaning out the combs, feeding the voracious queen? And what prompts her at three weeks of age to shrug off her nurse's togs and venture out into the world as a forager, a tireless gatherer of nectar and pollen, and the happenstance key to floral fecundity? What changes occur in the bee brain that might explain the dramatic career shift, with its concomitant capacity to fly a dozen miles a day and not get lost, and to dance the sororal dance that soundlessly booms to workmates the location of blossoms worth probing?

Robinson's team presented various threads of experimental evidence that a gene designated (why not) the foraging gene might be at the heart of the professional overhaul. Firstly, the scientists demonstrated that if they removed all the foraging bees from a hive and thereby forced some of the young nurse bees to assume breadwinning duties prematurely, the foraging gene flicked on abruptly inside the cells of the bees' beleaguered brains. Secondly, they showed that if they fed young bees sugar water laced with a chemical known to stimulate the activity of the foraging gene artificially, the sedentary cell dwellers suddenly started venturing outside, precociously prepared to gather ye rosebuds. Finally, if the researchers gave young bees another sort of stimulatory chemical that failed to activate the foraging gene, the bees remained hive-bound, a demonstration that not just any chemical kick would do the trick.

Through each evidentiary strand, and every corresponding control, still the discovery held. Unless the foraging gene blazed on, the bee didn't budge. A modest finding perhaps, but one chiseled and polished until it was the bees' knees.

Scientists demand evidence, and they are merciless toward a researcher who gives a PowerPoint presentation with feeble data. "It's a very aggressive, confrontational process," said Lucy Jones. "Conflict is part of the day-to-day reality of how science is done." I have heard scientists guffaw loudly during talks, when it was quite clear that the presenter wasn't telling a Werner Heisenberg joke. I have seen scientists under fire turn as pale as marzipan and start to quiver and almost spit, though I have never seen one cry onstage; and murders in the scientific community are surprisingly rare, although suicides, unfortunately, are not. The scientific hazing can give the enterprise a doctrinaire air, one intolerant of creativity, new ideas, anything that might upset the complacent status quo. It feeds the familiar $E = mc^2$ of the Hollywood scientist-hero, the lone genius battling an entrenched and blinkered theocracy with only his girlfriend to believe in him and remind him to bathe at least once a week. Now, it is true that when a pharmaceutical company has a best-selling drug at stake, company scientists can be suspiciously quick to dismiss studies showing a cheaper, competing product to be as good or better than the company's billion-dollar gravy boat. Even without the lure of big profits, research scientists often have egos that might best be measured in the astronomical unit known as the parsec; as a result, scientists may defend their research and their perspective long after the data have naysayed them. David Baltimore recalled an MIT scientist who died only within the last couple of years and who was one of the last remaining critics of the theory of the origin of the universe that is now almost universally accepted by astronomers and indeed the entire scientific community. "He didn't believe in the Big Bang," said Baltimore, "and he was in everybody's face about it."

Egos and academic mastodons notwithstanding, scientists are deeply skeptical when they hear amazing new results, and with good reason: many of these results are bad, are more awful than offal — a product that at least has a shot at fertilizing something better down the line. "Most of the time, when you get an amazing, counterintuitive result," said Michael Wigler of Cold Spring Harbor Lab, "it means you screwed up the experiment."

People have the mistaken impression that the great revolutions in the history of science overturned prevailing wisdom. In fact, most of the great ideas subsumed their predecessors, gulped them whole and got bigger in the act. Albert Einstein did not prove that Isaac Newton was wrong. Instead, he showed that Newton's theories of motion and gravity were incomplete, and that new equations were needed to explain the behavior of objects under extreme circumstances, such as when tiny particles travel at or near the speed of light. Einstein made the pi wider and lighter and more exotically scalloped in space and time. But for the workaday trajectories of Earth spinning around the sun, or a baseball barreling toward a bat, or a brand-new earring sliding down a drain, Newton's laws of motion still apply.

"The rules of science are quite strict," said the Berkeley astronomer Alex Filippenko. "I get messages every day from people who have ideas that sound interesting but that are terribly incomplete. I tell them, Look, you have to formulate your proposal much more coherently, in a way that explains not only the one new thing you're concerned with, but that is consistent with everything else we know, too. Any new, revolutionary idea has to explain the existing body of knowledge at least as well as the ideas we already accept."

On very rare occasions, scientists present a revolutionary idea in such a compelling, comprehensive, and vine-ripened form that even the skeptics are sold. One example is the famously brief paper in the April 1953 issue of the journal *Nature* by James Watson and Francis Crick, describing the incomparably uncluttered structure of deoxyribonucleic acid, or DNA. For years, many of the world's great geneticists were convinced that proteins, rather than nucleic acids, carried genetic information in the cell. Their reasoning was simple. Proteins are complex. They are the most complex molecules known in the cell. Genetic information seems pretty complex. Who better to bear the burden of complexity than the complex? On beholding the elegance of the double helix, however, and the smartness with which the four subunits of the twisting ladder paired with one another, and the ease with which one strand of the molecule might serve as a template for creating an entirely new copy of DNA to bequeath to a daughter cell, geneticists realized how the entire story of life could be told in its taciturn code.

Another legendary wowzer occurred at a geoscience meeting in the 1960s, when researchers offered evidence for plate tectonics, the theory that explains the origins of the ragged peaks and plunging canyons, the sputtering fumaroles and shimmering lava flows, and all the other Ansel Adams centerfolds that surround us. Lucy Jones's thesis adviser was at the meeting and told her how extraordinary the presentation was. "The evidence was so overwhelming, so compelling," she said, "that nobody could argue with it." Even more surprising, she added, "nobody wanted to."

Such *Rocky* triumphs, though, are extremely atypical. More often, scientists carp and cavil, demand better controls, offer a contrarian interpretation of the results, or write snide comments in the margins of a peer's manuscript. More often, science progresses fitfully, and individual experimental results are as modest as a bee's cerebrum. This is not an indictment against science. The power of science lies precisely in its willingness to attack a big problem by dividing it into many small pieces, its embrace of the unfairly maligned practice known as reductionism. At the same time, the piecemeal approach demands that scientists be circumspect to an often tedious degree and that they resist — no matter how much they are pushed by their university's public relations department or by desperate journalists — making more of the data than the data make of themselves. It would be cheating to do otherwise. It would be cheating to declare that science works by isolating variables, one colored peg at a time; and then to decide, when you've got a handsome little result, that, whaddya know, you're a holist at heart, and that Whitman had a point about the universe being in every blade of grass. The best scientists don't overreach or grandstand, at least not until they've retired into the armchair comforts of emeritus professorship, a time of life sometimes referred to as philosopause.

For working scientists, by contrast, all chairs are folding chairs: here today, tossed in the closet tomorrow. Scientists are accustomed to uncertainty, and to admitting how little they know. In fact, not only are they accustomed to uncertainty — they thrive on it. This is another of the core messages they'd like people to absorb, right down to their stem cells if possible: that science is an inherently uncertain enterprise, and that the uncertainty is, paradoxically, another source of its power. "We're out there looking for new patterns, new laws, new fundamentals, new *uncertainties*," said Andy Ingersoll, an astronomer at Caltech. "And as we're looking, and discovering new things, we're debating about what we see. We express our differences of opinion, sometimes strongly, until the public gets confused. Doesn't science know the answer to anything? Well, yes, eventually a consensus may be reached about a particular problem. But by then, we've already moved on to the next uncertainty, the next unknown. You don't linger." Ignorance is bliss, and always an excuse. "What motivates scientists is a lack of information rather than the presence of information," said Scott Strobel. Sometimes a consensus really is consensual, as it overwhelmingly is with Darwin's theory of evolution by natural selection (and more on this profoundly important organizing principle of biology, and the circus of manufactured tsuris that surrounds it, later), and as it firmly is in the case of global warming. For all the talk of "controversy," the great majority of climate scientists concur that average temperatures on Earth are climbing, and that some, if not all, of the rise is the result of human activity, notably the compulsive burning of combustible materials to power every aspect of contemporary life, including the need for more air-conditioning.

At other times, a scientific consensus amounts to little more than mass agnosticism. Take the question of whether chemical pollutants contribute to breast cancer. On the one hand, many industrial chemicals have been shown to cause breast tumors in lab animals; inherited factors fall short of explaining most human cases of the disease; and breast cancer rates vary significantly from nation to nation, all suggesting that environmental carcinogens somehow contribute to the malignancy. On the other hand, study after study seeking to link pesticides, power plants, or other specific environmental insults to human cancer have failed to reveal any convincing connection, leaving most scientists either skeptical or resolutely noncommittal about the contribution of chemical pollutants to breast cancer — much to activists' dismay.

"You don't want people to think that science is a joke, and that we don't know anything," said the Caltech astronomer Chuck Steidel, "but the truth is that the process of reaching a consensus is extremely messy and requires that a huge number of hurdles be overcome. Often, when results are presented to the general public, they're made out to be much more rock-solid than they are."

Science is uncertain because scientists really can't prove anything, irrefutably and beyond a neutrino of a doubt, and they don't even try. Instead, they try to rule out competing hypotheses, until the hypothesis they're entertaining is the likeliest explanation, within a very, very small margin of error — the tinier, the better. "Working scientists don't think of science as 'the truth,'" said Darcy Kelley. "They think of it as a way of *approximating* the truth." By accepting the proximate and provisional nature of what they're working on, scientists leave room for regular upgrades, which, unlike many upgrades to one's computer operating system, are nearly always an improvement on the previous model. For example, after scientists determined that DNA, rather than proteins, served as nature's preeminent guardian of genetic information, they began to see that DNA was not the sole guardian of the code of life, and almost certainly wasn't the original one. They gradually gained respect for RNA, the molecule they once dismissed as a mere bureaucrat paper-clipped between the imperial DNA that issues commands in the cell and the industrious proteins that do the cell's work without surcease. Scientists spied in RNA many talents that made it a likely ancestor of DNA, the primordial vessel of heredity and continuity back when life was new; only later did RNA cede its replicative and procreative role to the sturdier strands of DNA.

More recently, scientists have amassed evidence that some proteins, called prions, can act like DNA after all, replicating in the brains of mad cows and their unlucky human consumers. The discovery of prions and their infectious, photocopying potential earned a Nobel Prize for Stanley Prusiner in 1997.

None of these findings undermine the strength of the original Watson-Crick discovery. "Just because RNA and proteins can carry information in some circumstances doesn't detract from the centrality of DNA as the primary bearer of hereditary information," said David Baltimore. "As our concepts become more precise, more sophisticated, the absolutes become less absolute." In other words, by accepting that they can never *know* the truth but can only approximate it, scientists end up edging ever closer to the truth. The tonic surgery of chronic uncertainty.

For those outside the operating theater, however, all the quarreling, the hesitation, the emendations and annotations, can make science sound like a pair of summer sandals. Flip-flop, flip-flop! One minute they tell us to cut the fat, the next minute they're against the grains. Once they told us that the best thing to put on a burn was butter. Then they realized that in fact butter makes a burn spread; better use some ice instead. All women should take hormone replacement therapy from age fifty onward. All women should *stop* taking hormone therapy right now and never mention the subject again. Didn't scientists predict in the 1960s that a population bomb was about to explode, and that we'd all die of starvation or crowd rage? Now demographers in developed countries fret that women aren't breeding fast enough to restock the tax base and that nobody will be around to pay tomorrow's nursing home bills. Why should we believe anything scientists say? For that matter, why should we do anything that scientists suggest, like thinking about global climate change and the inevitable depletion of Earth's fossil fuels and adjusting our energy policies accordingly? That's what scientists say today. But if I hang on to my Hummer long enough, hey, maybe they'll decide that extravagant plumes of exhaust fumes are good for the environment after all!

This is one of science's bigger public relations problems. How do you convey the need for uncertainty in science, the crucial role it plays in nudging research forward and keeping standards high, without undermining its credibility? How can you avoid the temptations of dogmatism and certitude without risking irrelevance? "People need to understand that science is dynamic and that we do change our minds," said Dave Stevenson. "We have to. That's how science functions.

"Part of critical thinking," he added, "includes the understanding that science doesn't deal with absolutes. Nonetheless, we can make statements that are quite powerful and that have a high probability of being correct."

One trick to critical thinking is to contrast it with cynicism, which happens to be one of my most comfortable and least welcome of mental states. Cynics dismiss all offerings, sight unseen, data unmulled. Another drug that cures breast tumors in mice? Go tell it to Minnie. The fossil of a new dinosaur species disinterred? I can hear Stephen Jay Gould grumbling from the great beyond: Dinosaurs are a cliché. Preemptive cynicism may be rooted in insecurity, defensiveness, a gloomy disposition, or simple laziness; whatever its cause, it is useless.

Deborah Nolan of the University of California, Berkeley, encounters it constantly in her introductory statistics course — the slapdash bashing, the no-it-all choir. She confronts cynicism calmly and strives to replace it with hard-nosed thought. Each semester she'll present her students with newspaper stories that describe an array of medical, scientific, or sociological studies: Should victims of gunshot wounds be resuscitated by the paramedics in the ambulance, through drugs delivered intravenously, or is it better to wait until they get to the hospital? Does a surgeon perform better while listening to music in the operating room, or not? Does the mental well-being of a mother have a greater impact on her interaction with an infant, or with a toddler? Nolan will ask the students for their impressions of the articles. Regardless of the subject matter, or whether the students are majoring in science, the liberal arts, or hotel management, their initial response is the same: a synchronized sneer. You can't believe what you read in the newspapers, they'll insist. Nolan asks them what, precisely, they don't believe about the stories. They examine the articles again, this time with more care. Well, it's just . . . why *should* I believe it?

Nolan then shows them the original journal studies on which the newspaper stories were based, and she and the students begin, methodically, to pick the studies apart. They consider who the research subjects were, whether the participants were divided into two or multiple groups, the basis on which they were assigned to one group or another, and how the groups were compared. They discuss the strengths and limitations of the study, and why they think the researchers designed it as they did, and what the students might have done differently if they were running the study themselves. Enlightened now with this insider's intelligence, the students then reread the newspaper stories, to see if the reporters accurately conveyed the essence of the studies.

Most of the time, Nolan said, the students are impressed and appreciate that the reporters did their jobs after all, a change of heart that so surprised me I had her repeat the words slowly and clearly and right into my tape recorder.

More to the point, when the students come across an example of ineptitude, they can articulate why they feel dissatisfied. "They started off being highly skeptical of everything they read, without knowing quite why," she said. "But as critical thinkers, they could back up their comments and misgivings with precise descriptions of what was in the original study and what was omitted."

I also like Bess Ward's method for converting her students from cynical derision to clinical precision. Ward is a professor of geosciences at Princeton University, and every year she tells her students, Pick a worry, any worry. She has them pose a question about an everyday concern of theirs, a personal habit or indulgence or preferred food that they may have heard or read a negative report about. Their task is to figure out, Should I really worry, or not? How big a risk am I taking if I continue to eat or act as I do, and how does this risk compare to other risky behaviors that I freely or of necessity engage in? Or should I feel guilty about my little luxuries because they may be harming others, or are bad enough for the environment that I can't quite justify them?

"I tell them, choose something that you relate to and that may sometimes nag at you from the background of your mind. Drinking a lot of coffee, or taking birth control pills, or eating tuna sandwiches, or bungee jumping," she said. "The idea is, look at the evidence and do a risk assessment."

For most of these concerns, the basic data points, the worry wartlets, are accessible on the Internet. The Environmental Protection Agency's Web page, for example, offers so-called reference doses for virtually every toxic chemical you're likely to encounter — scientific estimates of how much of the chemical you can be exposed to without suffering harm. Here you will find the average concentration of mercury in an average Charlie tuna presented as milligrams of toxin per kilogram of fish. You will also find how many milligrams of mercury a person can safely ingest per kilogram of his or her own body weight before needing to worry about achiness, bleeding gums, swelling, blindness, coma, and, well, I think I'll just go with the arugula salad, thanks.

Or let's say you're fretting, as one of Ward's students did, over the relative riskiness of a weekly manicure. When you're in a nail salon, you're breathing in all the fumes from nail lacquers and the solvents that remove them, an ambient nosegay only slightly more sensual than that of the elephant facility at the National Zoo. But is obnoxious necessarily noxious? On the EPA Web page, you will discover that nail polish and polish remover contain toluene, a moderately toxic petroleum extract that also happens to be moderately volatile — i.e., it evaporates easily into the air you'll soon be breathing. The EPA also offers figures on toluene concentrations in different workplace settings, including nail salons. Elsewhere on the Internet, you can gather results from inhalation surveys to see how much air the average person breathes in over the course of an hour, which is about how long you'll spend on a task that is literally as thrilling as watching paint dry. After analyzing these and other statistics, you may conclude, as the young student did, that her weekly manicures are reasonably harmless, but that she wouldn't want to work ten-hour shifts in a nail salon and that maybe she should give really big tips to the women who do.

Another surprising barrier to thinking scientifically is that we often believe we already understand how many things work, especially simple things we were supposed to have learned in one of our formative, single-digit grades. Even absent specific exposure to this or that kiddie science problem via a parent, a camp counselor, or the Professor on *Gilligan's Island,* we develop an intuitive grasp of physical reality, a set of down-to-earth, seemingly sensible explanations for everyday phenomena: why it's hot in the summer and cold in the winter, or what's going on when we throw a ball into the air. Sometimes these intuitive concepts are so comfortably lodged in our brains that if that tossed ball were to become a cartoon piano and fall on our heads, we'd pick ourselves up like a dazed Wile E. Coyote, shake the twinkling phosphenes from our eyes, and go back to our same misguided schemes for catching the bleep-bleep Road Runner.

Susan Carey, a professor of cognitive neuroscience at Harvard, has explored the ways that our lovingly cultivated and often erroneous models of physical reality can subvert understanding and impede our capacity to learn. She uses as an example a ball that has been tossed into the air and then falls back to the ground. Say you draw a picture of this trajectory, she said, with a series of balls in a steep arc to represent the ball rising upward, at midpoint in the air, and coming down again. You then ask people to draw arrows showing what sort of forces they think are acting on the ball during its trajectory — their strength and direction. The vast majority of people look at the picture and draw big force arrows pointing up while the ball is headed skyward, and big arrows pointing downward while the ball is descending. A sizable fraction of respondents, recognizing that gravity is acting on the ball during its entire voyage, will add little arrows pointing down next to the big arrows pointing up for the ascent portion of the curve. For the ball at its zenith, many will draw a little up arrow and a little down arrow that effectively cancel each other out.

It makes sense, doesn't it? Ball going up, force arrows pointing up; ball going down, force arrows plunging earthward. In fact, it makes so much sense that people believed exactly this model of motion for hundreds of years. There's even a name for it — the impetus theory, the idea that when something is in motion, a force, an impetus, must be keeping it in motion. As reasonable and as obvious as this theory seems, however, it is wrong. True, there was an upward force exerted on the ball when it first was thrust into the air, compliments of the pitcher. But once the ball has been launched, once it is in midexcursion, there is no more upward force acting on it. Once the ball is in the air, the only force acting on it is gravity. All those arrows on the diagram should be pointing down. If there were no gravity to worry about, a ball tossed upward would keep sailing upward, no further encouragement necessary. This is one of Isaac Newton's many brilliant productions, the famed law of inertia: an object at rest tends to stay at rest, unless induced by the nudge of a police officer's stick to get up off the park bench, this isn't the Plaza Hotel, you know; while an object in motion tends to stay in motion unless a force is applied to stop it. Yet even though we have heard about the law of inertia, and have seen the movie showing what happens when a jealous computer clips an astronaut's tether in the weightlessness of space — there he go-o-o-es — still we have trouble applying the idea of inertia to something in motion, and still we draw diagrams of ascending balls with upthrusting arrows.

"People come to science learning with a coherent, rather systematic theory of mechanical phenomena, and it's usually a variant of impetus theory," said Carey. "And often, as they learn about Newtonian theory, force, momentum, inertia, pressure, they simply assimilate the new information into their preexisting concepts." She and other researchers have found that even among people who have had a year of college physics, a high proportion will explain the ball's trajectory in impetus terms. "They hadn't undergone a conceptual change," she said. "The intuitive concepts they started with still held sway."

Sometimes a piece of knowledge learned early can make a powerful impression, can become an intuitive understanding that is then summoned forth in a valiant effort to explain something else. For example, researchers have shown that many people, on being asked why it is warm and sunny in the summer and cold and sullen in the winter, attribute seasonality to the comparative distance between Earth and the sun. They begin by stating a fact picked up at some point in elementary or high school — that Earth's orbit around the sun is not a perfect circle, but an ellipse. They then explain that, when Earth is closest to the sun on its ovoid track, we have summer; and when it is farthest away, it's time for road salt.

Walter Lewin, a professor of physics at MIT, showed me a video of Harvard seniors being asked, at their commencement ceremony, to explain why we have seasons. Again and again the young men and women, cucumber-confident in their caps and gowns, explained it as a matter of Earth being farthest from the sun in winter and closest in summer. The respondents weren't all art history or English majors, either, but included a few physics and engineering students as well.

Lewin, who is Dutch and therefore gratuitously tall, has an Einsteinian froth of whitish hair, a loping, electric style, and a facial expression often tuned to an impish, resigned incredulity. "The misconceptions of high school," he said, "can dog you for the rest of your life."

It's true that Earth's orbit is elliptical, he said, but only modestly so. Yet when the students try to explain in a drawing how the shape of our planet's orbit causes the seasons, they invariably exaggerate the eccentricity of the ellipse into something with the contours of a Tic Tac. Now they have a visual representation of how they view the seasons. You see way out here, at the farther elliptical tip of the orbit? That's winter. You see this tip, where we're squeezing toward the sun? That's summer. "They fail to ask the question, If this were the case, why, then, is it winter in the Southern Hemisphere when it's summer in the North, and vice versa?" said Lewin. "They can't shake the image of the all-powerful ellipse from their minds."

As it happens, Earth is slightly *farther* from the sun in July than it is in December, yet none of this matters. Seasonality is the result, not of orbital geometry, but of Earth's tilt: the fact that the globe is spinning on an axis that is tipped over 23 degrees relative to the plane of Earth's migration around the sun. As a result, sometimes the Northern Hemisphere points toward the sun and is bathed in a comparatively stronger and more direct blast of heat and light, and everybody living between Caracas, Venezuela, and Wood Buffalo, Canada, is advised to wear plenty of sunscreen, long-sleeved clothing, a sombrero, and a canvas tarp. Six months later, when Earth is at the opposite end of its lazy-Susan revolution, the Northern Hemisphere is tipped away from the sun, and it's the Southern Hemisphere's time to get braised.

Again, most people know about Earth's tilt, if for no other reason than their childhood exposure to that obligatory household prop, the four-color globe, on which half the countries have long since been renamed, redrawn, and overtaken by a military junta, and which was rarely used except for the purposes of spinning it around on its notably slanted axis until it squealed. Because the spinning was understood to explain why we have days and nights, however, the angle of the rotation was as likely to be erroneously lumped together with the day-night kernel of kiddie wisdom as with any explanation for snow days and summer vacations.

Nor is it necessary that we learn our misinformation in childhood to hang on to it as a toddler would a small, shiny choking hazard. Whether sizing up new acquaintances or seizing on novel ideas, we remain forever at the mercy of our first impressions. We hear an explanation for something we hadn't been exposed to before, it sounds good and tastes better, and — you didn't just swallow that thing, did you? Cindy Lustig, a professor of psychology at the University of Michigan, recently demonstrated the ease with which our mind makes up its mind about new things. She gathered together forty-eight of the standard academic research subjects — undergraduate students — and instructed them to make an association between two related words, like "knee" and "bend" or "coffee" and "mug."

On a follow-up test, she asked her subjects to change the association, so that instead of answering the "knee" cue with "bend," the person was to reply "bone"; for the coffee prompt, "cup" rather than "mug." OK, time for lunch. Later that day, Lustig divided the group of subjects in two. Half were told to revert to the original association when confronted with the cue word. No problem: knee *bend,* coffee *mug.* The other group was asked to say whichever of their learned responses came to mind. Half of them would reply "bend" or "mug," and half "bone" or "cup." Good enough. Flip of the coin. Ah, but the next day, what then? When the random-answer subjects were again asked to say whatever response came to mind on hearing their cue words, a sizable majority conjured up their first tutorial, getting the bends, getting mugged. The earliest link, said Lustig, had become the brain's default setting.

Reporters know this tendency all too well, of the mind's readiness to make a quick connection and then seal it with an acrylic topcoat. I remember writing a story for the front page of the *New York Times* in 1991, about the spectacular discovery that we humans and other mammals have many hundreds of genes devoted to the production of odor receptors, the molecules studding the cells of our nasal passages that allow us to detect the thousands of aromas surrounding us. When I first heard the name of one of the smell researchers, Linda Buck, I immediately thought of another Linda with a similar surname, Linda Hunt, the New Jersey–born actress who won an Academy Award for playing a Chinese-Indonesian man. Well, both names are U-based, and you can *hunt* a *buck,* right? Ding-dong, connection made! Which is which? A wicked switch! I continued reporting the story. The hours flapped past. And when I finally got down to writing, I couldn't help but revert on cue to the earliest connection I'd made in the "Linda with the monosyllabic, rather bland last name" category, and I typed in Linda Hunt. Only at the last minute, right before the piece was to go to press, did I double-check the name against the journal article — and gasp at my error. Fortunately, I had time to make the change and save myself from prolonged humiliation. Linda Buck and her collaborator, Richard Axel, have since been awarded the Nobel Prize for their discovery, but there's still no Oscar in sight.

While simple facts like name spelling are easy to check and correct, it's much trickier to confront your preconceptions and misconceptions and to articulate how or why you conceive of something as you do. Your ideas may be vague. You're not sure where they came from. You feel stupid when you realize you're wrong, and you don't want to admit it, so you say, To hell with it, I'm no good at this, good-bye. Please don't do that. If you realize you might have put those up arrows on the ascending ball, too, or you weren't sure about the seasons, or you thought the lunar phases were the result of Earth's shadow being cast on the moon, rather than the real reason (that half the moon is always lit by the sun, and half is always dark, and that as the moon makes its month-long revolution around Earth we see different proportions of its light and dark sides), blame it on the brain and its insatiable greed, for picking up everything it comes upon and storing it in the nearest or most logical slot, which may not be right, but so what. That you have to be willing to make mistakes if you're going to get anywhere is true, and also a truism. Less familiar is the fun that you can have by dissecting the source of your misconceptions, and how, by doing so, you'll realize the errors are not stupid, that they have a reasonable or at least humorous provenance. Moreover, once you've recognized your intuitive constructs, you have a chance of amending, remodeling, or blowtorching them as needed, and replacing them with a closer approximation of science's approximate truths, now shining round you like freshly pressed coins.

2

Probabilities

For Whom the Bell Curves

AT THE START of each semester, Deborah Nolan teaches her elementary statistics students a basic, bilateral lesson in life: that it's really hard to look accidental on purpose; and, on the flip side of the same coin, that randomness can look suspiciously rigged. And what better way to prove her point than by flipping coins?

Nolan divides her class of sixty-five or so students into two groups. The members of one group are instructed to take a coin from their purse, pocket, or friendly neighbor, and to flip the coin one hundred times, recording the results of each toss on a sheet of paper. The other students are told to *imagine* tossing a coin one hundred times, and to write down what they think the outcome would be. After signing their work with an identifying mark known only to themselves, the students are to place the spreadsheets of heads and tails face-down on Nolan's desk.

Nolan then leaves the room, and the students start flipping coins and writing, or coining flips and writing. On returning, Nolan glances over the strings of one hundred Hs and Ts and declares each to be either real tossups or faked ones. Nolan is nearly always right, and the students, she said, are "aghast." They think she must have cheated. They think she peeked or had an informant. But she doesn't need to play Harriet the Spy. As it happens, true happenstance bears a distinctive stamp, and until you are familiar with its pattern, you are likely to think it is messier, more haphazard, than it is. Nolan knows what real randomness looks like, and she knows that it often makes people uncomfortable by not looking random *enough.*

In the real tossing of a coin, flick after flick, you will find many stretches of monotony, strings of five heads or seven tails in a row. Now, this is no big deal if you do it long enough and begin to realize that, in the course of one hundred or two hundred flips, clumping happens. Yet when we watch somebody flip a coin in shorter stretches, and especially if we have something riding on the outcome — who gets to choose the vacation destination, for example, or who has to remove the dead opossum from under the porch — we become very dubious when the coin starts repeating itself. *Six* tails? Where did you get that quarter from anyway? a Tom Stoppard play?* Let *me* try.

In their fantasy flippings, the students compensated for their inherent chariness of "too much coincidence" by frequent hopping back and forth, head to tail. In general, the act of jotting down a triplet would set off an alarm bell in the student's head, resulting in a deliberate change of face. "When I look at the fabricated coin tosses, the length of the longest run of heads or tails is way too short," said Nolan. "And overall, the number of switchbacks between heads and tails is way too high." People know there's a fifty-fifty chance for a given outcome with each toss, and they know that, on average, one hundred tosses will yield something close to fifty heads and fifty tails. OK, forty-eight tails, fifty-two heads, I can live with that. But six tails in a row?

"People want to apply the fifty-fifty rule over a very short period of time," said Nolan. "They have a skewed sense of probabilities, and they think the odds of getting multiple heads or tails in a row are much smaller than they are. In fact, the probability of getting four heads or four tails in a row is one in eight, so there's a pretty high chance of it happening." Nolan derived her figure by using the simple multiplication rule that applies to figuring out coin-flipping odds.† You have, of course, a 50 percent chance of tossing a head (or a tail) with each throw — in other words, a probability of 0.5. To calculate the odds of getting two heads in a row, you multiply the two odds together: 0.5 times 0.5, or 0.25 — a 25 percent chance that you, the penny pitcher, would see a pair of Lincolns. If you want to ratchet up the number of flips in your probability estimate, just keep multiplying. The prospect of seeing four

* Tom Stoppard's pleasurably unsettling comedy *Rosencrantz and Guildenstern Are Dead* opens with Rosencrantz and Guildenstern on the road to Elsinore, repeatedly flipping coins and getting heads every time.

† This multiplication rule only applies to calculating probabilities when each event in the sequence is independent of the other, as it is when you're tossing a coin. It could not be applied in cases where one event is likely to influence the other. For example, you can't calculate the likelihood of a man having a beard and mustache by multiplying the individual probabilities together, because, Abraham Lincoln notwithstanding, men with beards generally opt for mustaches as well.

heads emerge with four tosses is thus 0.5 quadrupled, which works out to a one-in-sixteen chance. But because we specified beforehand that we wanted to calculate the odds of seeing four heads *or* four tails, rather than four heads, period, we must add the two probabilities together, and one-in-sixteen plus one-in-sixteen is one in eight.* Granted, the odds of remaining one-sided decrease considerably with each additional toss. The likelihood of flipping six consecutive heads *or* tails is only about one in thirty-two, or 3 percent. This modest potential, though, applies to a single bout of a half-dozen flips. When you're flipping a coin one hundred times, the odds begin to add up, and so, too, do the clusters.

I tried Nolan's coin-tossing exercise myself several times, and over a dozen rounds of one hundred flips each, I never completed a set of one hundred without getting at least one string of six or seven heads or tails in a row, often more than one unbroken sextuplet per set, as well as many quintuplets and quartets. My record for monotony was nine heads in a row, which even now, knowing what I know and assuming a determination to outfox the instructor, I would feel queasy about including in a display of faux flipping.

Until they're schooled in the expansive possibilities of probability theory, Nolan's students regard the notion of randomness as a kind of nervous tic: sorry, sorry, can't stop twitching! Anything beyond this perpetual pinging and ponging, Abe and his monument, and what would you have? A pattern. From a pattern, it's a small step to assuming a point or a portent, and the next thing you know, some poor rabbit is forfeiting its foot to a key chain. "Because many people don't have a real feel for how likely it is for events to happen, they start to attribute hidden meaning to something that's random," said Nolan. "If they see a run of heads or tails beyond a certain length, they begin looking for reasons."

Here we find the basis for superstitiousness, she said. A chance occurrence occurs. Not knowing the odds behind it, we marvel, Now, really, what are the odds? Surely too tiny for chance!

Alan Guth, a physicist at MIT, described an example from his own family of how easily we turn the random into an omen. An uncle of his, who'd lived alone, had been found dead in his home, and a policeman had come to deliver the bad news to Guth's mother. While the officer was there, Guth's sister, who was traveling on business, happened to

* This additive rule requires that the two events be mutually exclusive, and again coin tossing fits the bill: with only a single penny on hand, you can't flip four heads and four tails simultaneously.

call. "My mother and sister were both shocked at the timing of the call, that it coincided with the policeman's visit, and the news of my uncle's death," said Guth. "They thought there had to be something telepathic about it." When Guth heard from his mother of this "miraculous" instance of kin-based telecommunion, he couldn't help but do some quick calculations. As a rule, his sister phoned their mother about once a week. She tended to call either first thing in the morning or in the evening, when she had a free moment and when her mother was likeliest to be around. The policeman had arrived at his mother's house at about 5:00 P.M., and, because there were several solemn orders of business to discuss, his visit had lasted more than an hour, possibly two.

All factors considered, Guth said to me, the odds of his sister calling while the policeman was on-site were on par with flipping five heads or tails in a row. "This is not what I would consider a highly improbable event," said Guth. Lucky, yes, given his mother's need for comfort from a loved one, but nothing for which the telepathy option need be considered.

The more one knows about probabilities, the less amazing the most woo-woo coincidences become. My mother told me an amusing story about an acquaintance of hers whose fate, over a six-month period, had seemed linked to her own as though by an idle Pan. The acquaintance was, appropriately for our purposes, an old math professor of hers. Week after week, my parents kept running into him somewhere on Manhattan's sprawling cultural turnpike — an off-Broadway play, a free piano recital, a Bergman movie, the Monet *Water Lilies* room at the Museum of Modern Art. The first few times, my mother and her professor chortled awkwardly over the similarities of their taste. Soon, they were content to nod vaguely from across the room. The coup de graceless came a few months later, in July, and in another country. My parents were strolling along the boulevard St.-Michel on their first trip to Paris, when who should they see but the good professor, sitting at a café. Judging by the way he held his newspaper ostentatiously in front of his face, my mother knew he had spotted them first.

Had my mother been of a superstitious bent, she might have thought the universe was trying to tell her something. ("Your professor hates you!") She is, however, one of the least superstitious people I know, and she understood that (a) those who like Monet like French art; (b) Paris is famous for its world-class collection of French art; (c) "April in Paris" sounds romantic, but "An American in Paris" sounds like July; and (d) an outdoor café is the best place to while away many hours not drinking a cup of cold espresso, not smoking the lit Gauloise in the ashtray, and not really reading the *Herald-Tribune.*

John Littlewood, a renowned mathematician at the University of Cambridge, formalized the apparent intrusion of the supernatural into ordinary life as a kind of natural law, which he called "Littlewood's Law of Miracles." He defined a "miracle" as many people might: a one-in-a-million event to which we accord real significance when it occurs. By his law, such "miracles" arise in anyone's life at an average of once a month. Here's how Littlewood explained it: You are out and about and barraged by the world for some eight hours a day. You see and hear things happening at a rate of maybe one per second, amounting to 30,000 or so "events" a day, or a million per month. The vast majority of events you barely notice, but every so often, from the great stream of happenings, you are treated to a marvel: the pianist at the bar starts playing a song you'd just been thinking of, or you pass the window of a pawnshop and see the heirloom ring that had been stolen from your apartment eighteen months ago. Yes, life is full of miracles, minor, major, middling C. It's called "not being in a persistent vegetative state" and "having a life span longer than a click beetle's."

And because there is nothing more miraculous than birth, Deborah Nolan also likes to wow her new students with the famous birthday game. I'll bet you, she says, that at least two people in this room have the same birthday. The sixty-five people glance around at one another and see nothing close to a year's offering of days represented, and they're dubious. Nolan starts at one end of the classroom, asks the student her birthday, writes it on the blackboard, moves to the next, and jots likewise, and pretty soon, yup, a duplicate emerges. How can that be, the students wonder, with less than 20 percent of 365 on hand to choose from (or 366 if you want to be leap-year sure of it)? First, Nolan reminds them of what they're talking about — not the odds of matching a particular birthday, but of finding a match, *any* match, somewhere in their classroom sample. She then has them think about the problem from the other direction: What are the odds of them not finding a match? That figure, she demonstrates, falls rapidly as they proceed. Each time a new birth date is added to the list, another day is dinged from the possible 365 that could subsequently be cited without a match. Yet each time the next person is about to announce a birthday, the pool the student theoretically will pick from remains what it always was — 365. One number is shrinking, in other words, while the other remains the same, and because the odds here are calculated on the basis of comparing (through multiplication and division) the initial fixed set of possible options with an ever diminishing set of permissible ones, the probability of finding no birthday match in a group of sixty-five plunges rapidly to below 1 percent. Of course, the prediction is only a probability, not a guarantee. For all its abstract and counterintuitive texture, however, the statistic proves itself time and again in Nolan's classroom a dexterous gauge of reality.

If you're not looking for such a high degree of confidence, she adds, but are willing to settle for a fifty-fifty probability of finding a shared birthday in a gathering, the necessary number of participants accordingly can be cut to twenty-three. Throw a couple of dozen people together at a cocktail party, in other words, and you have a slightly better than even chance that two of them will be birth-date mates, who, if they discover the fact, will likely exclaim over the coincidence and segue to a discussion of astrology. Or, if their birthday happens to be February 16, and they're talking to me at this imaginary cocktail party, they will hear of the many other date mates who preceded them — Susan the San Francisco photographer, who always brought her golden Labradors on assignment; Frank the Atlanta businessman, who briefly sublet my apartment and whooped it up at the neighborhood tiki bar; Michelle, my brother's girlfriend; and, first but ever least, Robbie, a high school boyfriend of mine, who was cute and smart and studiously mean. Maybe it was his rising sign, or something his poor mother ate.

Through exercises like Birthday Buddies, Nolan's students begin to see the world as both surprisingly predictable and full of surprises. It is a place where small numbers can take on grand airs and seem, on first pass, more meaningful than they are: how could a meager number like 23 possibly perform like 365 without some sort of cosmic motivational speaker prodding it from behind?

It is also a venue large enough for rarities to become regulars, where so many millions of lottery tickets have been sold that ridiculous patterns emerge. A sixty-year-old Australian man buys a Lotto ticket before leaving for vacation, worries that he bought the wrong sort of ticket, and asks a friend back in Sydney to buy another, then frets on returning home that his friend fumbled the request and so decides to spring for a third entry — and ends up with three winning tickets in hand. A woman in Milwaukee, Wisconsin, responds to her husband's hankering for an expensive experimental airplane kit, Sure, honey, go ahead and splurge "when you win the lottery," just as her father had won the state's $2.7 million Megabucks jackpot a dozen years earlier; her husband takes the suggestion seriously and scores a $2.5 million

Megabucks prize himself. Or officials for a large multistate Powerball lottery drawing become suspicious when 110 players scattered among the 29 participating states come in to claim second-place prizes, rather than the 4 or 5 such winners expected from the drawing. But each of the 110 petitioners had guessed 5 out of the 6 Powerball numbers correctly, and each was entitled to anywhere from $100,000 to $500,000 apiece, depending on the initial bet. Behind the startling outbreak of good fortune was a fortune cookie. All the second-place winners had based their choice on the 6 digits they'd seen on the little slip of paper tucked inside a Chinese fortune cookie, a fortune that, like the cellophane-wrapped bill brightener that held it, had been produced in bulk at the Wonton Food factory in New York.

Most of us are not accustomed to a probabilistic mindset, and instead approach life with a personalized blend of sensations, convictions, desires, and intuitions. Our gut is certainly a significant piece of property. The gastrointestinal tract measures about thirty feet from throat to rump and accounts for 10 to 15 percent of one's body weight — but its physical dimensions are nothing compared to its metaphoric value, as the source of our cherished "instincts." We meet new people, we size them up and get a "gut feel" for what they're like, and we contrast them with others in our acquaintance until we find the closest fit. Ah, now we've got them sussed, trussed, and mounted. Now we can safely nap. If our gut instinct happens to clash with logic, probability, or evidence, guess which claimant wins?

Jonathan Koehler of the University of Texas admits that he is not always a popular guest at a wedding. He sits at the ceremony and listens to the giddy couple exchange vows of permanent devotion, passion, and respect. He hears the toasts attesting to the unmistakable rightness of the match, how anybody who knows this man and this woman could tell from the start that the union was "meant to be" and is "like no other," and he thinks, Hmm, I've been to four weddings in the past year. Who's it going to be, then: Zack and Jenny? Sam and Brianna? Brad and Briana? Or Adam and Hermione, now lip-locked so protractedly before me? Which two of these four pairs of besotted newlyweds will end up carrying botulinum-tipped spears into divorce court ten years hence? After all, minor fluctuations notwithstanding, the American divorce rate has been remarkably stable at 50 percent for nearly half a century.

Koehler is friendly and chatty and sometimes shares his musings with other wedding guests. They look at him as though he had belched, or speculated on the correlation between the size of the bride's brassiere and that of the groom's paycheck.

"They find it repugnant to talk statistics at a wedding," he said. "They want to know how I can say such a thing. Why, you don't know anything about this couple! Just look at how happy they are, how deeply in love, how overjoyed their families are. True enough — but I know general frequency statistics. I also know that every couple gets married with kisses and toasts and high hopes, so these details shouldn't affect the probabilities we assign to them. Until you tell me something outside the norm, something diagnostic that has been shown to affect one's probability of divorce — for example, both partners being over the age of thirty-five, which is known to lower the probability of divorce — I'll assume the normal statistical risk applies." Koehler, who has the slight build and dark, floppy hair of Michael J. Fox, insists he's not a "cynical, bitter little man" or a self-satisfied bachelor: to the contrary, he recently got married himself. He's simply accustomed to viewing the world as an extravaganza of sample spaces.

"People don't tend to pay attention to the background information, the sample space," he said. "They take the foreground information without context, and they accept it at face value."

And while full frontal credulity may be the lubricant of matrimony, he said, at other times it helps to look at the big-sky backdrop. More than once Koehler has calmed a jittery passenger seated next to him on an airplane by quoting probabilities. You would have to fly on a commercial airline every day for 18,000 years, he tells them, before your chances of being in a crash would exceed 50 percent. You want to know what 18,000 years looks like? Think "twice as far back as the dawn of agriculture."

Koehler has also examined the errors that people make in deciding how to invest their money. In one study, he and his colleague Molly Mercer showed subjects mockups of advertisements for mutual funds. To the first group they displayed an ad from a small company with a phenomenal track record. It operated only two funds, but each consistently outshone a benchmark market index like Standard & Poor's. Now it was starting up a third fund: Wanna invest? The next set of subjects was treated to an ad from a large mutual fund company, which mentioned that it ran thirty funds and then showed the results of the two funds that "killed" the market index; it, too, was seeking investors for a new fund. Yet another group saw a pitch from the same large company, again attempting to entice investors to a new fund by highlighting the lavish returns on its two star funds, but this time with no reference to the many other, and presumably far less impressive, money sinks in its corporate portfolio.

Koehler and Mercer found that subjects generally were impressed by the small company's results and voiced a willingness to buy into their latest start-up fund. They were equally unimpressed by the big company with thirty funds. "People recognized that, Oh, you're showing me only the best two out of thirty, and they said, Sorry, not interested," said Koehler. But when confronted with ad number three, from the big company that boasts of its two knockouts while omitting any reference to its baseline operations, subjects again fell prey to the lure of the fabulous foreground. They greeted it with the same enthusiasm accorded the small company.

"From a mathematical standpoint, the fund from the investment group that's two-for-two is a much better risk and is much likelier to outperform the market than is that of a group that's two-for-question-mark," said Koehler. "But people often forget to ask, What's the question mark here? They're not thinking about the sample space."

Unfortunately for us poor hayseeds seeking a place to plant our paychecks, real-life advertisements for mutual funds are not legally obliged to divulge their losers and thus they rarely do. Even the advice of "experts" may not enhance our prospects. "We got the same pattern of responses to our ads," Koehler said, "whether we asked undergraduates or professional investors."

Koehler conceded that it's not easy to think about a sample space, the background context, the teeming multitudes beyond the home team in front of you. "We're not hard-wired to think probabilistically," he said. "We're hard-wired to respond to life subjectively, empathetically, and on the fly, which may be a generous impulse in some cases, but at other times it clouds our judgment and is flat-out wrong." One approach he takes to encourage a quantitative mindset is applying it right where subjectivity has the greatest stranglehold on sense: our people skills. He uses exercises like the notorious Linda Problem. Students are given a paragraph describing a hypothetical character named Linda, who is said to be a thirty-year-old American woman who majored in philosophy, graduated with high honors, and has been active in the nuclear freeze and antidiscrimination movements.

Following that tapas of a biography are eight statements, which the readers are asked to rank in order of probability that they apply to Linda. Among them: Linda is a bank teller; Linda is a feminist; Linda is married and has two children; Linda lives in a university town; Linda is a feminist and a bank teller.

Time and again, Koehler said, readers think they know Linda. She's a feminist — that they rank high. And she probably lives in a university town. The married-with-kids part, who can say, so that gets a listing somewhere in the middle. But a bank teller? That description doesn't sound like Linda at all, and it earns an average ranking way at the bottom of the stack. She could, however, be a feminist *and* a bank teller, couldn't she? Readers assign the composite declaration a higher probability than bank teller alone. "Almost ninety percent of people do this," said Koehler. "They argue, she's definitely not a bank teller, but she could easily be a bank teller and a feminist. At least that's got some of Linda in there. That seems to be the way people think about probability."

There is, of course, a higher probability of Linda being a bank teller than a bank teller and a feminist. In order to be a bank teller and a feminist, she must be a bank teller; and the unconditional probability of one event occurring — in this case, bank tellerdom — is always going to be greater than the conditional conjunction of that event plus a second event — bank tellerdom and a familiarity with the works of Simone de Beauvoir and Gerda Lerner.

Yet even as people accept that Linda might be a feminist bank teller, they feel uncomfortable thinking of Linda's overall prospect of being a bank teller, period. Some might think that to use the job description alone negates, misrepresents, or shortchanges an essential aspect of her being, just as I've felt compelled to qualify my answer whenever people have asked what my father did for a living: he was a machinist at Otis Elevator Company, I say, but he was also an artist who made intricate pen-and-ink drawings, i.e., he was no Archie Bunker. Alternatively, people might be unconsciously fleshing out the statement "Linda is a bank teller" with a clause, "but she is not a feminist," to place it in direct contrast to the statement "Linda is a feminist and a bank teller."

However understandable and folksy may be the urge to rank the conditional above the unconditional premise in Likely Lines about Linda, it is incorrect, and when Koehler's students realize the error of their weighs, they feel foolish at first, and then eager to try the trick on family and friends, and finally liberated. Where else can they apply their newfound wisdom, their awareness of how important it is to consider background?

Nowhere is the utility of sample-space tracing more obvious than when interpreting the results of a medical test. As a number of studies have revealed, doctors are not always skilled at estimating probabilities or putting a test result in proper context, which means that patients may be sent into paroxysms of anxiety, soul-searching, and planning of funeral choreography unnecessarily, or at least prematurely.

Let's take as an illustrative but purely hypothetical example the following scenario. You're at the doctor's office for routine maintenance, and you happen to notice a sign advertising the monthly special: an AIDS test that is described as "95 percent accurate." You are not in any of the standard high-risk groups for the disease — though you did have crab lice back in college — but as a conscientious citizen and aspiring hypochondriac, you decide to roll up your sleeve and get screened. A week later, the receptionist from the temp agency who's been filling in for your doctor's phlebotomist calls with grim news: you tested positive. You feel the blood abandon your head and reconvene around your plantar warts. You can't speak. The receptionist mumbles how sorry she is, and how she loved Tom Hanks in *Philadelphia.* How sorry should *you* be, especially since you've never forgiven Hanks for *The Man with One Red Shoe*? The test is "95 percent accurate." You came out positive. Assuming the results weren't caused by a major mechanical screwup like a swapping of test tubes or charts at the laboratory, there's a 95 percent chance you're infected with the AIDS virus, right?

Unbate your breath. Even if it was your vital fluid that yielded the positive result, the real odds are much, much smaller than 95 percent that you are genuinely HIV-positive. In the lively Port Said of the free market, the definition of a test's accuracy can vary depending on the needs and temperament of its parent pharmaceutical company, but in general this figure would mean the following: on the one hand, the test will accurately detect the human immunodeficiency virus in 95 percent of those who have it but will fail to catch 5 percent of those infected; on the other hand, it will correctly rate as negative 95 percent of all noncarriers, but — and here's where your comfort food comes in — it will mistakenly generate a positive result for 5 percent of uninfected patients. Why should you find solace in a puny false-positive figure like 5 percent? Because the potential pool, the sample space, embodied in that figure is formidable. In the United States, HIV infection remains relatively rare, afflicting about 1 in 350 people. Taking a more population-worthy slant on the problem, that means in a random group of 100,000 Americans, some 285 will be HIV-positive, and 99,715 not. Yet if we screened all 100,000 with our AIDS test, what would we expect? The assay would accurately pick up 271 of the 285 viral carriers; but it would slap a fallacious writ of panic on some 4,986 noncarriers. To calculate the odds that a positive result means you are actually infected, you divide the total number of true positives you'd expect in your sample space (271) by the total number of positives overall — false (4,986) and true (271) together. Slice 271 by 5,257, and you end up with a probability of 5 percent. The gist of that calamitous phone call, then, amounts to the flip figure of your initial fears: there is a 95 percent chance you're virus-*free.**

None of this is to suggest that estimating probabilities in most real-world settings is easy, or that you should start second-guessing medical advice by running your test results through a two-way ANOVA statistical analysis. Still, it never hurts to ask some simple questions, such as, How common is this illness or condition in the general population? In other words, What is the size of the sample space I'm up against? This question is particularly useful in trying to get a reasonable sense of a "risk factor," or of one's "relative risk" compared to Max and Bryanna Populi. For example, five bad sunburns before the age of fifteen is said to double your odds of developing malignant melanoma. How awful! A few lousy days at Camp Minnehaha spent extracting oar splinters from your palms and taking group lanyard lessons under the full noonday sun, and you can raise your risk of contracting a potentially deadly skin cancer by *100 percent*? Yes, but as it happens, melanoma is quite rare, afflicting only 1.5 percent of the U.S. population; so even with the legacy of your childhood stir-fries, and assuming no other risk elevators like a family history of the disease, you're still talking about a lifetime risk below 4 percent. By all means, watch out for the appearance of new skin moles, particularly those shaped like raisins, Rorschach blots, or the literary caricatures of David Levine; and make sure that you and your loved ones are fully shellacked in sunblock prior to opening the window shades; but putting your dermatologist's pager on speed dial is surely going too far.

You might also want to ask your doctor about the published rates of false negatives and false positives for a given assay, and whether the measure of those accuracy statistics is itself accurate. Most health care professionals, despite their descriptor, are far more concerned with diagnosing and treating *illness* than they are in minimizing the number of false alarms their screens may activate among the healthy. As they see it, it's worse to miss a real case of a disease than to spot what initially looks like trouble and then find out, whew, you're fine after all. Yet for you the medical consumer, the devastating impact of a false positive, however brief its tenure, can feel like an illness, so if there's any way to

* I must emphasize that the "95 percent accuracy figure" bandied above is strictly hypothetical, and that the true accuracy rate for today's HIV tests is much better, greater than 99.9 percent. Nevertheless, the use of medical assays and "routine screens" is rising sharply, and many of them suffer from distressingly high and decidedly nonhypothetical rates of false positives. *Caveat patiens.*

combat it with an estimate like the one for our hypothetical AIDS test, fire away.

Another way to feel more comfortable around quantitative reasoning is to try some at home, starting with a fun exercise that I'll call, until somebody stops me, the Fermi flex, after the great Italian physicist Enrico Fermi. In addition to being one of the giants of twentieth-century science, Fermi was a leader of the Manhattan Project during World War II, an assignment that for some reason had its stressful moments. To fortify morale and remyelinate the frayed nerves of his fellow bomb makers, Fermi would throw out quirky mental challenges. How many piano tuners are there in Chicago? he might ask, or, How many pounds of food do you eat in a year? As Fermi saw it, a good physicist, or any good thinker, should be able to devise an ad hoc, stepwise scheme for attacking virtually any problem and coming up with an answer that lies within the vaunted terrain known as "an order of magnitude." In other words, you shouldn't have to multiply or divide your estimate by a factor of ten or more to embrace the real answer. If the real answer is 5,400, you should be able to get an estimate in the range from 1,000 through 9,999; if the answer is 33,000, your Fermi-approved margin extends from 10,000 through 99,999.

Flexible enough, but how can you even begin to approximate the dimensions of an obscure trade like piano tuning in a city with which you have only the barest of airport hub acquaintance? In his admirable book *Fear of Physics*, the fearless physicist Lawrence Krauss shows the way. Chicago is one of the nation's largest cities, he says, which means its population must be up in the multimillion range, but not the 8 million of America's urban heavyweight, New York. Let's give it 4 million. How many households does that amount to? Say four people per dwelling, or some 1 million households. Think about the rate of piano ownership among your acquaintances: maybe 10 percent of the homes you know? So we've got roughly 100,000 Chicago pianos in need of occasional tune-ups. What's "occasional"? Once a year seems like a reasonable guess, at a fee of, say, $75 to $100 per tune-up. Now consider how many pianos a full-time piano tuner must tune to stay solvent. Maybe 2 a day, 10 a week, 400 to 500 a year? So we divide 100,000 by 400 or 500. All conjectures hazarded, we might expect to find a labor force of 200 to 250 pulling strings somewhere in the fabled birthplace of the skyscraper, the well-tailored gangster, and a bland, eponymously named rock band from the 1970s. By the order of his majesty's order of magnitude, Krauss writes, "this estimate, obtained quickly, tells us that we would be surprised to find less than about 100 or more than about 1,000 tuners." No need for shock therapy: the actual answer is about 150.

My turn. I decided I'd try estimating the number of school buses in my county in Maryland, Montgomery, which extends from the border of Washington, D.C., at the southern edge up to points north near Baltimore. Mainly I was curious about how many buses sit idle during the county's vast number of "snow" days, which in this delusional plow-averse state are declared, not on the basis of verifiable accumulations of the white, fluffy substance called "snow," but rather on the premonition of snow as determined by a single factor: before venturing outside, you must put on something called "a coat."

In any event, how many of those cheery yellow child chariots can Montgomery County claim? From my obsessive scrutiny of election results every November, I happen to know that the county has about 500,000 registered voters. I also know that, given its proximity to our nation's capital, the region is politically plugged in and has a high rate of voter registration, maybe 70 percent, among eligible citizens. So I'd estimate the adult population to be around 650,000, or about 300,000 potential pairs. How many of these adult pairs are between the ages of twenty-five and fifty-five, the demographic likely to have school-age children? Let's say 150,000. And let's say that half of them have children, the most popular number being 2 per couple, with maybe 1.5 of those offspring in school. That gives us 110,000 kids in the Montgomery County school system. Some of those children are in private schools; others live close enough to walk or sniffle piteously enough to get driven. Let's cut the bused population in half, to 55,000. How many little scholars can you pack into one vehicle? Maybe 50? So that brings us down to about 1,100. But before we rest on our guesstimate, we must recall that school buses barrel through multiple routes each morning, which is why the wretched teenagers living next door to me have to be up and out the door to catch their bus by 7:15, while my elementary-school daughter gets to leave seventy minutes later. Assuming two routes a day per vehicle, we might wager that there are some 550 school buses in the Montgomery County public school system. Or at least somewhere between 100 and 1,000.

Consulting the Web page for the Montgomery County school system, I find that it owns about 250 school buses, half of my predicted sum, but still well within an order of magnitude of it. True, you could conclude that I might have saved myself the trouble by consulting the Internet to begin with; but I appreciated the exercise, the thinking through of the different parts of the puzzle — the number of fecund adults that might surround me, the likelihood of them acting out their fecundity, how many kids are in my daughter's cohort of standardized test–takers, and so forth. Through regular sessions of Fermi flexing, you get a better sense of how the world looks and how the pieces fit together. And while learning to admit that you don't know something is a worthy skill in its own right, better still if you can rally an algorithm to relieve your ignorance. If you're talking to a coworker who tells you his goal is to jog the equivalent of once around the Earth, and you realize with some embarrassment that you don't know or can't recall the circumference of the Earth, and you don't like this pompous coworker enough to give him the satisfaction of asking, Oh, and how far might that be? you can do a quickie estimate. Think about some geo-detail you do know — say, the duration and destination of a very long flight. My husband recently flew nonstop from New York to Singapore aboard Singapore Air; and though he slept for most of the eighteen-hour journey, he did manage to collect goodies like a cute hot-water bottle and a pair of booties with antiskid strips on the bottom. Singapore is very far from America's eastern seaboard, just about halfway around the globe, I'd guess. Jets average some 500 to 600 miles per hour. So 9,000, 11,000, miles to Singapore, and double that for a round-the-world belt of 18,000 to 22,000 miles. The circumference of the Earth, in fact, is 24,902 miles at the equator (or 40,076 kilometers to most earthlings, including those who live at the equator). Our frequent-flier-derived answer, then, is well within the Fermi order of magnitude mandate. Yet jet-setting is one thing; literal globetrotting quite another. Glancing at the generous circumference of your colleague's waistline, which does not bespeak a natural athlete's physique, you smile broadly and wish him Godspeed. Why, a random act of quantitative reasoning has even made you appear kind.

For all the power of quantitative reasoning and probabilistic analysis, Mark Twain, as ever, had a point about statistics: damn, can they lie. One of the finest and funniest popular science books ever written is the 1954 classic *How to Lie with Statistics,* by Darrell Huff, on the theme of how the experts are doing exactly that to you every day. Take the much-bandied and seemingly redoubtable term "statistically significant." Call a result "statistically significant," and it sounds as though there's no arguing the point. "Even some scientists and physicians have been brainwashed into thinking that the magic phrase is the answer to everything," said Alvan Feinstein, a professor of medicine and epidemiology at the Yale University School of Medicine. But what does "statistically significant" signify? Although definitions vary depending on who's bandying, the unadorned phrase generally means that the correlation you the scientist have hit upon — an association between a particular genetic mutation and a disease, for instance — has a probability value, or p value, of 5 percent, which in turn means that there is, at most, a 5 percent chance that your patent-pending correlation was due to chance alone. In other words, there is a 95 percent chance that you are onto something. A "p = 0.05" is the minimum passing grade that, according to scientific convention, renders a result "statistically significant" and eligible for submission to at least a sprinkling of the 20,000 or so research journals published worldwide. Yet consider how easy it is to beat this degree of significance to a senseless blubber. The hypothetical AIDS test discussed earlier would have a p value of 0.05; that's what its "95 percent accuracy" rate is all about. The outcome? A pool of false positives big enough to do laps in. For this reason, many scientists don't feel comfortable with such a lax measure of confidence, and they won't publish until their p values have a couple more zeros to the right of the dot, and the odds of the result being a mere fluke pretty much equal to their chance of, say, winning the Nobel Prize. Twice.

Another slippery statistics term that has found its way into popular usage, and political abusage, is "average." As in: the average tax refund from the president's tax cut program will be $1,500. That sounds pretty decent, until you discover that the statistical "average" doesn't mean the "usual amount" of rebate that the "usual sort" of American family can expect to see. The statistical average, which is also known as the norm, is the statistical *mean,* a number you get by adding up all your quantities and dividing the sum by the number of data points — in this case, the grand total of tax refunds divided by the number of rebate checks cut. The problem with such calculations is how readily they can be skewed by, for example, the inclusion of a few colossal givebacks. If twenty families living on Creston Avenue in the Bronx receive tax refunds of anywhere from $100 to $300 per household, but a family with a floor-through on Manhattan's Gramercy Park gets an IRS mash note worth $70,000, the "average" refund for those twenty-one families would be about $3,500. Gee, thonx, said the Bronx. I feel richer already. Do you mind if I give a Bronx cheer?

A much more revealing data point would be the *median* tax cut, the value you'd see if you laid each of the twenty-one rebate checks in a row from feeblest to fattest and looked at the figure on the midpoint refund — the eleventh check. It would be about $200, a far truer measure of what the average Jones in our sample received than is the obfuscating "average." These days, given the growing gulch between extreme wealth and ordinary income in our country, financial matters often are best explored as medians rather than as averages or norms. If you include the wealth of a few Bill Gateses and Warren Buffetts in any calculus of "income norms," you'll make the whole population look comfortably flush, even as the great majority of families earn considerably less than your stated average or indeed what they might need to cover their monthly Visa bill.

Yet means and medians are not always so mismatched. Many times, they congregate closely beneath the comfortable shade of the celebrated parasol we know as the bell curve. This essential scientific principle unfortunately took on a neocon connotation in the mid-1990s, when Charles Murray and Richard Herrnstein adopted it as the title for their best-selling book about race and IQ. But Bell Curve: The Concept is much deeper and more illuminating than Bell Curve: The Tract. It's extraordinary how much of the world settles into a bell curve when you sally forth to size up its parts. If you were to go into a field of daisies, and measure the heights of, say, three hundred flowers, and mark those heights on a graph, you'd find a few shorties on the left end of the chart, and a few gangly overreachers on the right, but the great majority would amass in the midrange, and the contours of your distribution plot would, yes, ring a bell. The same for measurements you might make of the daisies' leaves, or of the diameter of the yellow centers. You'd have a few outlier examples of any given feature — stubby leaves, moon pie faces — but most would cluster around a central value that, whether you figured it as the mean or the median, would pretty much define the average dimensions of this most fetchingly normative of floral ambassadors.

In her class, Deborah Nolan also brings the bell curve to life by playing tailor to her students. "I take many different measurements of them, height, shoulder width, the distance from the shoulder to elbow, the elbow to the fingertips, the distance from the pinkie to the thumb." Plotting the results of each tally on the blackboard for her five or six dozen students, she shows them how nature adores a good hump.

The same bell curve contour would define the results of coin-flipping bouts. If you performed 1,000 bouts of 100 coin flips, you'd have a sprinkling of really skewed ratios of, say, 71 heads and 29 tails, or even a freakish 80-something tails, teen-something heads, but the great bulk would be in the neighborhood of fifty heads and fifty tails.

Finding the contours of a normal distribution for a given problem is part of what science is all about. What's your mean value, and how do you know when you've got it? If you're trying to figure out the average alcohol consumption among students at a local college, how many people must you interview to feel confident you haven't inadvertently sampled a few too many frat boys, liars, or Seventh-day Adventists? When do you know you've amassed a large enough sample that the midpoint of your bell curve has meaning, that it captures the representative slice of reality you're after? You don't want to end up like the three statisticians out on a duck hunt: the first one fired a shot that sailed six inches over the duck, the second fired a shot six inches under the duck, and the third one exulted, "We got it!" The rules for determining the statistical soundness of a sample size are complex and depend on the particulars of the problem, but a couple of tenets generally apply: the sample should be as large as is practically and economically possible; and once the sample population is settled on, the net should be as finely meshed as you can make it. Nothing tarnishes the credibility of a sample like the desire to be sampled, which is why the results of a sex survey of the readers of *Maxim* magazine may be far less revealing than any of the garments on the females displayed therein. A good pollster will hound and rehound the very people who least want to cooperate.

The fact that so many things in life, from the length of a human pinkie to a roll of the dice, conform to a bell curve pattern of data points says something fundamental, if potentially dispiriting, about life: that it's much easier to be ordinary — that is, to dwell somewhere within the normal distribution of whatever category you're measuring — than to be outstanding (or, for that matter, grossly inadequate). Parents want each of their offspring to be what Gertrude Stein is purported to have called "an immortal something or other"; and inspirational spots on public television always feature children dreaming of being great successes — the next Thomas Edison, a world-famous chef, the first astronaut on Mars. Yet distribution theory reveals that values cluster around midpoints, and that mediocrity loves company. As a result, the only way for most children to be "outstanding," "genius material," or even merely "gifted and talented" is to redefine your terms ("of course you're extraordinary: there's never been anybody in the history of the human race with precisely your DNA!"), inflate your grades, or dump your rankings altogether.

Bell curves aren't cast in bronze, and their midpoints can be coaxed over a bit in a preferred direction, usually gradually, sometimes dramatically. With a few changes in public health practices, for example, like pumping sewage out of town instead of slopping it out the window, and encouraging doctors to scrub their hands between patients, the average life span in the United States nearly doubled between the mid-1800s and the mid-1900s. In another twentieth-century great leap upward, the American-born-and-fed children of immigrants soon towered over their parents, pushing the two bulges of the bell curves for height — one for women, one for men — rightward by several inches. Average IQ scores also have risen in the past half century, for reasons that remain unclear.

Whichever way a bell curve swings, there is always a big fat greedy bulge somewhere, sucking up the bulk of the population. Indeed, the pull of the bell's bulge is so relentless that it's been given its own term: regression to the mean. By this principle, the extraordinary tends to lose its edge over time. If two unusually tall parents have a child, the child is likely to be taller than average, but slightly shorter than his or her same-sex parent; the child, in other words, will regress toward the mean. Why should this be so? Because the parents reached their imposing stature through a combination of genetics and a series of small happenstances during development that all shook out in favor of added verticality; and though they may pass along genes that generally enhance height, the chance settings that accentuated their loftiness will be reset to zero with the new generation and are unlikely to reposition themselves as a series of pluses once again. It can happen, but the odds are against it, just as they are against a mother flipping five heads in a row, handing the coin to her daughter, and having her daughter promptly repeat the trick. While population averages in height or intelligence may advance over time, regression to the mean serves as a counterweight, a stabilizing trend that helps keep cockiness in check.

John Allen Paulos proposes that regression to the mean could explain the legendary *Sports Illustrated* jinx: the long-standing observation that quite often, after an athlete appears on the cover of *Sports Illustrated,* that person goes into decline, fumbling the ball, botching the serve, assaulting the fans. Such unstellar turns could result from the pressures of fame, or a superstition subsumed into self-fulfilling prophecy, but Paulos thinks otherwise. "When do you appear on the cover of *Sports Illustrated*? When you've done extraordinarily well for a period of time and are at the top of your game," he said. "By implication, you're not going to be able to maintain your outlier status for very much longer." You are going to start regressing, however slightly, back toward the mean streets of the mean.

The same might be said for many a miracle cure in the annals of alternative medicine. People often resort to alternative therapies when they have been ill for some time, and have failed to find relief in a mainstream medicine chest. They are at their wits' end, desperate for relief. A friend recommends bee pollen, or shark cartilage, or powdered bear carbuncle, and they decide to give it a swallow. A week later, they're largely healed; after two, enzealed. Why didn't their physician recommend bear carbuncle in the first place? Was it because the pharmaceutical industry can't patent or profit from it and so hasn't distributed educational literature and free samples? Or was the doctor too narrow-minded to consider a therapy that looks like the sort of thing you can order through the back pages of the *Utne Reader*? Perhaps. Or perhaps the cure had nothing to do with the ingested novelty item, and instead represented another instance of regression to the mean. After many weeks precariously poised on the outlier tail of illness, people slip back into the comfortable lap of health, the physiological norm that our immune system grants us most of the time and that we take for granted until it is gone.

That people readily attribute a spontaneous recovery to some bold move, some agency, on their part, demonstrates the human desire to feel in control of one's destiny, yes. But it also underscores our readiness to conflate correlation with causation, which brings us to yet another way in which we may be snookered by statistics. Just because two traits or events are frequently found in the same package doesn't mean that one is responsible for the other. Sometimes the independence of oft linked items is easy to discern. In Sweden, many people are blond and blue-eyed, but obviously the Viking coolness of their gaze is not what blanched their hair, or vice versa. At other times, conjoined traits seem more portentously causal, but one must take great care before sketching out the flowchart. For example, many high school dropouts smoke cigarettes. Among adults in the United States, 35 percent of those who never finished high school are regular smokers, compared to 14 percent of those with a college degree. But does one characteristic in this correlation cause the other, and if so who does what to whom? Do high school dropouts smoke at two and a half times the rate of college graduates because they left school before learning just how bad the habit is? Do they smoke comparatively more because they're likelier to be in dead-end jobs that make them depressed, and nicotine, as a compound that both stimulates and relaxes, is just the sort of double-edged drug depressives crave? Or did their addiction to cigarettes prompt them to drop out in the first place — to get a job to support an increasingly expensive habit, or to escape the chronic censure of their teachers? Or are dropping out of high school and smoking cigarettes useful as signs of sedition, to advertise one's hostility toward society? Or are dropping out and smoking signs of submission, to advertise one's fealty to a gang?

Drawing causal arrows from one behavior or outcome to another is often fraught with danger, but that doesn't stop people from trying. In *How to Lie with Statistics*, Darrell Huff cites an example from a Sunday supplement called "This Week," in which an editor answered a reader's question about the effect that going to college has on one's odds of remaining unmarried. "If you're a woman, it skyrockets your chances of becoming an old maid," the editor replied. "But if you're a man, it has the opposite effect — it minimizes your chances of staying a bachelor." The editor then quoted from a Cornell University study of 1,500 "typical middle-aged college graduates," in which 93 percent of the men were married, compared to 83 percent for the general population, while only 65 percent of the women were married. "Spinsters were relatively three times as numerous among college graduates as among women of the general population," the editor ominously concluded. The lesson for the 1950s gal was clear: going to college, like getting fat or contracting a mild case of polio, can seriously diminish one's romantic opportunities. Boys do not wed the bookish coed.

Hold your Miss Havishams, huffed the progressive-spirited Darrell. Before we breezily turn a correlation into an open-and-shut case of cause-and-effect, who's to say that all those "old maids" in the Cornell survey pined to get married in the first place? They could very well have seen college as a way to escape matrimony and gain economic independence. For that matter, if college-bound women are relatively more single-minded than other women to begin with, who knows what impact their university experience may have had on them; perhaps even fewer of the Cornell coterie would have gotten married if they hadn't gone to college. All these possibilities are equally valid conclusions, said Huff. "That is, guesses."

Those who are statistically sophisticated can, if they choose, squeeze a number set until it squeals "Ninety-six Tears." Sir Richard Peto, an epidemiologist at the University of Oxford, made this point absurdly clear when the editors at *The Lancet* asked him to perform additional statistical analyses on a landmark report he and his colleagues had just submitted to the British medical journal. In their study, the researchers showed that heart attack victims had a comparatively better chance of surviving if they were given aspirin within a few hours of the attack. *The Lancet* editors wanted the epidemiologists to break the data down into subgroups, to see whether different patients might benefit more or less from aspirin depending on their age, previous health status, or other characteristics. Sir Richard balked. He knew that if you fiddled with and whittled down your numbers long enough, all sorts of spurious connections might arise through chance alone. The editors insisted. Finally, Peto relented, and gave them the subsidiary calculations they desired — but only on condition that they include in the publication one statistical "link" he'd uncovered that would drive home the need to regard the whole subgroup massage exercise with appropriate skepticism. Welcome back to the zodiac. Aspirin may be a lifesaver for heart attack victims born under ten of the twelve astrological signs, Peto wrote, but for those who happen to be a Libra or a Gemini, so sorry, the drug appears to be worthless. (Note to Libras and Geminis with current or suspected cardiac activity: consult your doctor, astrologer, or local cable company about whether "salicylic acid" might be a better choice for you; but under no circumstances should you contact Dr. Peto, who is a Taurus.)

In a similar bid to demonstrate the dangers of crackpot correlations, Sherman Silber, a reproductive surgeon in St. Louis, and two colleagues published the results of their willfully whimsical fishing expedition through a database of twenty-eight infertility patients. They used a computer program to identify any traits whatsoever that might link those women who had succeeded in becoming pregnant. Bless my speculum, what have we here: those patients whose last names began with the letters G, Y, or N were significantly more likely to end up bearing a child than were their less auspiciously surnamed peers. After admitting to a certain amount of ego gratification at the coincidence, Dr. Silber warned that many a "statistically significant" correlation in the scientific and medical literature may be just as specious as his game of GYN-ecology, but that few, unfortunately, will be as "patently ridiculous" and thus as easy to defrock.

If it's hard for the workaday doctor or researcher to recognize every sham correlation that might pop up on PubMed, none of us can escape the occasional hoodwink. And as tempting as it might be to defend yourself proactively by damning all statistics indiscriminately, Frederick Mosteller, a statistician at MIT, had a point when he said, "It is easy to lie with statistics, but it is easier to lie without them." Nevertheless, there are some steps you can take to, as Huff put it, "talk back to a statistic." Among the biggies recommended by many scientists is to ask a simple question: Does the figure, finding, or correlation make sense, that is, accord with what you know of objective reality? "You have to look at the biological plausibility," said James L. Mills, chief of the pediatric epidemiology section of the National Institute of Child Health and Human Development. "A lot of findings that don't withstand the test of time didn't really make any sense in the first place."

I once reported on an astonishing discovery from the world of primatology: In the typical social grouping of chimpanzees, with multiple adult males living with multiple adult females, and manic multiple mating on the scale of the old Manhattan swingers' haunt Plato's Retreat, it seemed that the resident males were often wasting their time, Darwinically speaking. Yes, they were consorting with the resident females, over and over again, but DNA analysis seemed to indicate that, despite the males' exertions, half the baby chimpanzees in a given group had been sired by fathers other than the resident studs. How could this be? The discovery roiled the close-knit but competitive community of chimpanzee researchers. Over decades of fieldwork, devoted ape gapers from Jane Goodall onward had seen virtually no evidence of extratroop cavorting, of females sneaking off for liaisons with nonresident males.

The short answer to "How could it be?" is "Oops." As another team of researchers determined a year later, the finding that defied biological plausibility turned out to be erroneous, the regrettable outcome of suboptimal genetic samples crossed with misleading statistical comparisons of the chimpanzees' DNA. On reanalyzing the DNA fingerprints, the primatologists brought molecular evidence into alignment with field studies and showed the resident male chimpanzees to be the true father figures to whatever hairy bairns would dare gambol among them.

Once again, the abiding scientific verity proved apt: when confronted with an astonishing result, cache a kernel of doubt until the finding has been independently verified, preferably by an old rival of the researcher who had hoped to do anything but.

Other questions to ask of a statistic include: Who discovered you? Was it an interested party with an economic, emotional, or political stake in the outcome? Pharmaceutical companies had abundant incentive to promote so-called hormone replacement therapy as a cure for anything that frails you, and for a few years in the 1990s huge numbers of women were convinced that the benefits of drugs like Premarin in keeping their hearts hale, their spines straight, and their collagen bouncy far outweighed any small, added risk of breast cancer the hormones might bring. But when a reasonably impartial jury, the Women's Health Initiative, tackled the worthiness of the hormones on a nationwide scale, they found the risks dwarfed the benefits, that in fact the benefits were almost negligible. Unfortunately, most drugs are not subject to a similar degree of federally financed scrutiny. The pharmaceutical industry pays for most of its own safety and efficacy trials, and, yes, many instances of corporate chicanery or negligence have surfaced over the years: warnings about the dangers of the painkiller Vioxx ignored, evidence that some antidepressants may raise the risk of suicide among adolescents suppressed. Still, your best bet is to ask where a statistic comes from, and whether it has been verified by an impartial source.

As mentioned earlier, you should also seek to put a statistic in context and bring key background facts to the fore. If you hear that the incidence of a childhood cancer rose by 50 percent between last year and this, take a look at the numbers for the preceding five years. Childhood cancers are always devastating, but thankfully even the commonest members of the perverse class — leukemia, for example, or neuroblastoma — are still quite rare. With rare diseases, a few extra cases can make a huge difference in rates. Look at how the figures fluctuate over time. If there's been a slow but steady rise in incidence over a decade, then a report warning of the trend merits attention. But for an erratic zigging and zagging, random misfortune is as likely an explanation for a bad year as anything else.

Above all, remember that numbers are not mystical, infallible, or always pure of heart. Many people say they hate being treated as "just another statistic." Well, a statistic is never "just" a statistic, either. It's the product of a human mind, a human judgment call, human imagination, human bias, human weakness. Learning to think quantitatively helps one surmount a tendency to accept a quantity without quibble or qualification. A young relative of mine recently took the SAT and scored 1,300 out of 1,600. My family obviously has known her for years, but now we had a quantity by which we could really peg her to the board: she's pretty smart, but not flagrantly smart. A few months later, without the aid of a tutor or a Stanley Kaplan course, she took the SAT again, and scored 1,410. Phew! She's not just pretty smart, she's extremely smart.

The Scholastic Aptitude Test may be a wholly hominid invention, written by a small cabal of elders for untold throngs of youngers, but we treated it as though it offered cosmic truth. And when it presented two different versions of that truth, we did what any loving family would do, and called the first figure a liar.

3

Calibration

Playing with Scales

Of the seven deadly sins, the one with perhaps the most diverse menu of antivenins is the sin of pride. Need a quick infusion of humility? Climb to a scenic overlook in the mountain range of your choice and gaze out over the vast cashmere accordion of earthscape, the repeating pleats swelling and dipping silently into the far horizon without even deigning to disdain you. Or try the star-spangled bowl of a desert sky at night and consider that, as teeming as the proscenium above may seem to your naked gape, you are seeing only about 2,500 of the 300 billion stars in our Milky Way — and that there are maybe 100 billion other star-studded galaxies in our universe besides, beyond your unaided view. A visit to a cemetery also does the trick: no, not one of those poignant churchyards tucked beside a James Renwick, Jr., cathedral, where the gravestones are few, slate, and safely antique; but a place like the Montefiore Cemetery complex in Queens, in which my grandmother, two of her siblings, and maybe another 150,000 of the recently deceased are buried, and which sprawls for several hundred acres just off the Long Island Expressway.

Yet of the many humbling tonics to which I have willingly or incidentally been exposed, perhaps the most effective was also the most humble. Not long ago, I revisited the old Bronx neighborhood where I spent my childhood, and I was overcome with an embarrassing case of existential angst. It wasn't that the neighborhood had changed terribly much. True, the apartment building where we lived had been torn down and replaced by a parking lot, but many of the surrounding prewar buildings were still standing, as grimily well intentioned as ever. Instead, what distressed me was how minor and compressed everything seemed, how much shorter the distances between the touchstones of my formative years were in reality than in memory. The geography of my childhood had been momentous, every block a continent, every ordinary excursion my own private odyssey. The weekly pilgrimage to the Garden Bakery for a loaf of challah or seedless rye, and maybe a black-and-white cookie if Lady Luck had chatted up my mother beforehand? Surely we're talking a half mile or more! No. The bakery is gone, but the corner remains, a mere two blocks from my home. The daily trek to P.S. 28, my elementary school, along a slalom of uphills and downhills, switchbacks and sinister intersections, and the dauntingly long stretch at the end where a gang of girls had assaulted me and stolen my brand-new purse? Four and a half blocks.

Obviously my sense of scale had been out of whack and off the map, a puerile version of Saul Steinberg's often imitated Manhattanite's view of the world. I'd felt overwhelmed in childhood by every detail of my microhabitat, and so I'd exaggerated the physical dimensions of my surroundings to match their emotional might. Now that I could size up the neighborhood through the pitilessly polished lens of adulthood, I realized how slight my all had been, how badly I had misjudged the distance between any two points. It wasn't my fault. I was a kid, and children are by nature preternaturally alert to the particulars of the niche into which they've been thrust. But the experience offered a graphic example of how often we humans stumble over our scales. Throughout history, people have wildly misjudged distances, proportions, comparisons, the bead of being. We non–Native American Americans owe our presence in and possession of the New World to that colossal navigational blunder called the "Enterprise of the Indies," Christopher Columbus's attempt to reach the Far East by sailing west. Maps traditionally are centered on the land most beloved by the mapmaker — Jerusalem to the medieval illuminist, country of birth or current employer to the cartographers of today. By all appearances, we have evolved to view life on a human scale, to concern ourselves almost exclusively with the rhythms of hours, days, seasons, years, and with objects that we can readily see, touch, and count on, because those are what we have to work with, those are the ambient utensils with which we must build our lives.

Yet the vital pacemakers and proportions of daily life are entirely incidental. Consider, for example, that satisfying quantity, the handful. We humans are able to glance at groupings of up to about five objects together and instantly know, without counting, the quantity, a skill thought to be a legacy of the five fingers with which we've always grabbed at treasures like ripe blueberries (or better yet, chocolate-covered blueberries) and against which we could then evaluate the magnitude of the plucked bounty. Yes, we have ten fingers, but we're a strongly handed species, about 90 percent of us right-handed, and we do most of our grasping with that favored fivesome. It's remarkable how difficult it is to see a cluster of, say, seven or eight objects and recognize them as such without going through the tedium of counting — unless, that is, they're arranged in tidy subgroups of five or fewer. Our sense of time, too, reflects our everyday experiences. The basic unit of ordinary time, the second, corresponds remarkably closely to the two most basic rhythms of life: the time it takes to fill our lungs with a breath and the duration of a single healthy heartbeat.

Because our solar system formed when a great mass of gas, dust, and rock began collapsing in on itself (a subject we'll take up in some detail later), and because gravitational condensation causes bodies to start spinning like those amazing vertigo-proof figure skaters, all the planets rotate on their axes at greater or lesser speeds. Earth happens to be spinning at a rotational speed that takes just about twenty-four hours to complete (23.934 hours, to be exact). As Annie Dillard said, "How we spend our days is, of course, how we spend our lives," and the boundaries of those days are the accidental bounty, a literal spinoff, of gravity. In fact, Earth's dervish dancing has been gradually slowing down, largely as a result of the tidal tugging of our tagalong moon. Early on, Earth completed a twirl in only ten hours, and even as recently as 620 million years ago a day was done in 21.9 hours, nightmarish notions for those of us already inclined to whine about deadlines and sleep deprivation.

Location is everything, and it was ours during the birth of the solar system that granted us our annum. Earth sails around its orbit of more than half a billion miles at 66,600 miles per hour because of its distance relative to the gravitational master, the sun. Venus, by contrast, is 26 million miles closer to the sun than we are, which means that (a) its orbit is shorter than ours; (b) the comparatively greater gravitational pull of the sun prompts Venus to dash through each lap at a heightened pace (78,400 miles per hour); and (c) a year there lasts only 226 Earth days, another unpleasant thought for book writers with contracts to fulfill. And let's not dwell on that solar toady of a planet named after the Roman god with feathers on his shoes, where a "year" lasts less than three months.

What little visceral sense we have of history tends to be based on the average human life span of some three-score and ten. Any interval greater than a century in either direction blurs in our mental calendar into an ameboid abstraction. I've known for most of my life that my ancestor Silas Angier fought in the Revolutionary War, but until recently I had no idea how many generations lay between him and me. When people would ask me, in light of my surname, whether I am French, I'd reply, Not lately, and I'd explain that the Angier family came to America from England in the seventeenth century; and while I was at it, I'd throw in a reference to my heraldic vinculum to our nation's founding. "In fact, my great-great-great, great-great" — a rapid waving of the hand backward through air and space-time — "great-great and so forth grandfather Silas Angier fought in the Revolutionary War." Wow, they'd say. Is there something wrong with your hand?

In the course of writing an essay about Fitzwilliam, New Hampshire, however, the town in which Silas and many other Angiers are buried, I had another of those back-to-the-Bronx moments, an embarrassed recognition of my distorted sense of scale. By going through town records, I determined that the greats between Silas and me were not so great after all, that they could be sized up handily with a finger to spare. He of the musket, breeches, and tricornered hat, the fellow born six years before Thomas Jefferson, was merely my great-great-great-great-grandfather. Contrary to myth, time doesn't fly particularly fast when you're dead.

Kings and assorted other highnesses often believed their personal parts to be of sufficiently divine proportions to merit adoption as standard units of measurement. The Roman emperor Charlemagne declared in the ninth century that the length of *his* foot would henceforth be *the* foot; and by that measure the emperor, who was said to be of good but not towering physical stature, could boast of standing seven feet tall. Three centuries later, the British monarch King Henry I decreed that the yard would be equal to the distance from his nose to the tip of the middle finger of his extended arm. The ever roaming Romans devised the concept of the mile as the distance a man can cover in 1,000 full-stride, manly paces; "mile" comes from the Latin term *milia passuum,* or thousand paces.

All such measurements were gradually standardized beginning in the Renaissance and continuing into the twentieth century. And while I am an ardent partisan of the metric system that has been embraced by all scientists and by virtually every nation save ours, I admit that there is nothing particularly fundamental about most of the metric units. They are not based on essential properties of atoms, or light, or gravity. (With one notable exception: the metric of temperature, degrees Celsius, is derived from critical phases of a cosmically abundant molecule without which we could not exist — water. The temperature at which water freezes is given the coveted slot of naught as 0 degrees Celsius, while the boiling point of water is designated as 100 degrees Celsius.) Regardless of origin, the metric system is defensible for its base-ten beauty, its ease of arpeggioing up and down the keyboard. How many millimeters in a centimeter, centimeters in a meter, meters in a kilometer? That's 10, 100, and 1,000 respectively. How many inches in a foot and a yard, how many yards in a mile? 12, 36, 1,760. Gee. Tough choice about which system we should be teaching our kids. So why does my daughter still have to learn both? When are we going to give up our inches, take out our miles, and toss them all on one last Fahrenheit fire? I have a sneaking but wholly unsubstantiated suspicion that the real block-and-tackle to American metrification is the American football field, and the hallowed quantum of the ten-yard line.*

Metric or otherwise, our anthropocentric sense of scale can impede our comprehension of the cosmos, indeed of virtually every science apart from the psychology of our distorted sense of scale. Thus, the scientists I interviewed were unanimous in their conviction that people would benefit enormously from a better grasp of nature's true dimensions: the length and breadth and tenure of the visible universe, the timeline of life on earth, the sublime spaciousness that persists even down to the imperceptible atom. Talk about the size of the cell, they said, and of the cell's citizens, the proteins, the hormones, the compressed coil of genes cloistered away in the nucleus. And what of the pirates that invade the cell: How big is yersinia, bacterial bearer of plague, compared to a white blood cell that yearns to knock it offstage? Remember the viruses: Where might Ebola weigh in? And how many of any could dance on a pin?

Frankly, I can't imagine a happier assignment than to talk about scales, especially because I don't have to step on any of them and then start pushing them around the bathroom floor until I find the best spot. Sometimes just knowing how the things you can't see compare to the things you can't miss is the better part of understanding. Moreover, practicing scales in nonhuman keys can have the salubrious effect of forcing you to question who's normal and who alien. "In my field of particle physics, the notion of time is essential, but we deal with times that are vastly different from everyday human concepts," said Robert

* Following the lead of my nation's educational system, I will alternate between metric and the old British units throughout the book.

Jaffe of MIT. "We deal with things like the time it takes light to cross a proton, on the order of 10 to the minus 24 seconds." In other words, a trillionth of a trillionth of a second. "People say, that's ridiculous, how can you be dealing with such ephemera," he said. "But the sense of alienation that people bring to the subject is a result of an anthropocentric concept of time that is in fact the real oddball here. Our perception of time is very unusual and hard to find in other systems of physics. It's easy to find extremely short time scales, like those that apply to many subatomic particles, and it's easy to find extremely long time scales, like those that pertain to the universe and to very stable particles, but it's very unusual to find scales like hours, days, and years. Our quirky concept of time has to do with the celestial mechanics of our solar system, and of the fact that we're poised between the energy scale of gravity and the world of nuclear forces."

To play any scales beyond our pedestrian ones, to talk of celestial harmonics or quantum dynamics, you need scientific notation, otherwise known as the powers of ten. The power of this notation has almost but not quite infiltrated popular culture, thanks in good part to Philip and Phylis Morrison's best-selling book *Powers of Ten.* But scientific notation deserves even greater magnitudes of fame, for it is both lovely and useful, like a fine old oak table with claw feet and spare leaves for when company comes. It's called powers of ten because you're asking, How many times do you have to multiply your figure by ten to get to where you're going? Ten times ten, or 10^2, is 100; ten times ten times ten, or 10^3, is a thousand. Add another power of ten to that string, and you've got 10^4, or 10,000. Scientific notation allows you to write perversely large numbers in compact form, and to manipulate them with the sort of ease rarely encountered beyond the privacy of your microwave oven. As of late 2006, for example, the U.S. national debt stood at $8.5 trillion. You can write that out in long form, as 8,500,000,000,000, and almost feel the red ink flowing from your veins. Alternatively, you can translate the quantity into scientific notation, by putting a decimal point immediately after your leftmost digit, and counting rightward to find your power of ten, or exponent. With a figure like 8.5×10^{12}, you won't feel nearly so overwhelmed, and may even come to think of such sums as reasonable and rational, at which point you'll be qualified to run the Office of Management and Budget.

To gain a quick grip on things by way of scientific notation, it helps to memorize those superscripts that correspond to numbers you know. A thousand with its three zeros is 10^3, a hundred thousand 10^5, a million

10^6, a billion 10^9, a trillion 10^{12}, a googol 10^{100}, a Google a search engine and transitive verb, and Gogol a nineteenth-century Russian novelist. You can see, then, why "exponential growth" is so pushy. The exponent of a billion may be only three more than that for a million, but that cute little three means, I raise you a thousandfold, dear.

Scientific notation works just as well for the furtive as for the discursive, although in this case you're talking about powers of one-tenth rather than powers of ten. One-tenth of one-tenth is one-hundredth, written as 10^{-2}; one-tenth of one-hundredth is one-thousandth, or 10^{-3}. Keep biting the right-handed bit of Alice's toadstool. Down you go, you're a fractionated Italianate family. You're milli — a thousandth, 10^{-3}; or micro — a millionth, 10^{-6}; or nano — a billionth, 10^{-9}; or pico — a trillionth, 10^{-12}; or femto — a millionth of a billionth, 10^{-15}.

Now we can start to examine a world that stretches beyond the realm of ordinary accountability. What happens, for starters, in subsections of seconds? In a tenth of a second, we find the proverbial "blink of an eye," for that's how long the act takes. In a hundredth of a second, a hummingbird can beat its wings once, and it is by the grace of this hyperbolic wing-flinging that these birds can hover like helicopters to sup in midair.

A millisecond, 10^{-3} seconds, is the time it takes a typical camera strobe to flash. Five-thousandths of a second is also the time it takes the *Bolitoglossa rufescens,* a Mexican salamander that resembles a blade of grass and that owns one of the fastest tongues in nature, to extrude its mauve sling and snag its prey.

In one microsecond, 10^{-6} seconds, nerves can send a message from that pain in your neck to your brain. On the same scale, we can illuminate the vast difference between the speed of light and that of sound: in one microsecond, a beam of light can barrel down the length of three of our metric-resistant football fields, while a sound wave can barely traverse the width of a human hair.

Yes, time is fleeting, so make every second and every partitioned second count, including nanoseconds, or billionths of a second, or 10^{-9} seconds. Your ordinary computer certainly does. In a nanosecond, the time it takes you to complete one hundred-millionth of an eye blink, a standard microprocessor can perform a simple operation: adding together two numbers, say, or flagging that questionable travel-and-expenses figure on your tax return.

The fastest computers perform their calculations in picoseconds, or trillionths of a second, that is, 10^{-12} seconds. If you could observe the intimate behavior of the water molecules in your lukewarm bottle of Dasani, you would see that every three picoseconds or so, the weak chemical links that hold adjacent water molecules together dissolve and reform again, a shimmering glimpse of the tentative nature of even the most carefully marketed products.

Ephemera, however, are all relative. When physicists, with the aid of giant particle accelerators, manage to generate traces of a subatomic splinter called a heavy quark, the particle persists for a picosecond before it decays adieu. Granted, a trillionth of a second may not immediately conjure Methuselah or Strom Thurmond to mind, but Dr. Jaffe observed that the quark fully deserves its classification among physicists as a long-lived, "stable" particle. During its picosecond on deck, the quark completes a trillion, or 10^{12}, extremely tiny orbits. By contrast, said Jaffe, our seemingly indomitable Earth has completed a mere 5 times 10^9 orbits around the sun in its 5 billion years of existence, and is expected to tally up maybe another 10 billion laps before the solar system crumples and dies. "That brings us up to 15 times 10^9 orbits, considerably fewer than 10^{12}," said Jaffe. "In a very real sense, then, our solar system is far less stable" than particles like the heavy quark. The shackles of "our personal, anthropocentric conception of time," said Jaffe, "make it hard for us to understand the vastness of the stability that these particles embody."

Scaling down to an even less momentous moment, we greet the attosecond, a billionth of a billionth of a second, or 10^{-18} seconds. The briefest events that scientists can clock, as opposed to calculate, are measured in attoseconds. It takes an electron twenty-four attoseconds to complete a single orbit around a hydrogen atom — a voyage that the electron makes about 40,000 trillion times per second. There are more attoseconds in a single minute than there have been minutes since the birth of the universe.

Still, physicists keep coming back to the nicking of time. In the 1990s, they inducted two new temporal units into the official lexicon, which are worth knowing for their appellations alone: the zeptosecond, or 10^{-21} seconds, and the yoctosecond, or 10^{-24} seconds. The briskest time span recognized to date is the chronon, or Planck time, and it lasts about 5×10^{-44} seconds. This is the time it takes light to travel what could be the shortest possible slice of space, the Planck length, the size of one of the hypothetical "strings" that some physicists say lie at the base of all matter and force in the universe. Chronons and strings remain more in the realms of mathematics and philosophy than empirical reality, however; and no one knows what would happen if we shaved our numbers further, and took a long gambol on a really short Planck.

The universe, though, doesn't only like to cut things short; it also opts for the sagging saga approach, dictating thick volumes of time that are nearly as unfathomable as *Finnegans Wake.*

Consider Earth time, which really is a Joycean "riverrun, past Eve and Adam's." If you had all the time in the world, what would you have? Creationists, scanning the pages of Genesis, Galatians, and other biblical sources, and counting up the "begat"s, bellow, "Six thousand years!" But the creationists' clock is — what's the word for "off by six orders of magnitude"? — cuckoo. There are one or two otherwise productive geologists who believe the biblical story of creation and insist that Earth really is young but that God has given it the illusion of great antiquity — but that's out of more than 100,000 geoscientists working in the United States alone. No, had you world enough and time, you'd have 4.5 billion years, for it was that long ago that Earth, and the other planets of the solar system, condensed from the flattened Frisbee of rock and dust surrounding the newborn sun. Now, on first mull, 4.5 billion years doesn't sound excessive, decrepit, or particularly awe-inspiring. After all, if you added up the birthdays of every human alive today, assuming a median age of twenty-six, you'd have about 170 billion years.

Yet 4.5 billion years stretched end to end, as they have been, lend Earth an extraordinary degree of flexibility, have made it a place where nearly everything is possible, the comical mandatory, the provisional a familiar pest who never misses a party. Over 4.5 billion years, seas and savannas have swapped places; Earth's magnetic poles have flipped and flopped and flipped again; glaciers have gripped nearly the entire globe in a snowman's nelson; and sumptuous tropical forests of towering club mosses and ginkgo trees, millipedes as long as men are tall, and dragonflies with a falcon's wingspan have stretched from Antarctica and Australia up through Europe and the Americas. Oh, yes, it can be almost impossible to think in geologic time, even for geologists.

"I look at time differently now that I am forty-six than I did when I was twenty, and I will look at it differently again when I am seventy-five," said the geologist Kip Hodges. "But none of this is going to put me in a position where I can understand 500 million or 650 million years, let alone 4.5 billion years."

In an effort to convey the great girth of Earth time, geologists who regularly communicate with the laity have conceived a wide assortment of metaphors and colorful visual aids, often involving long skeins of knotted yarn or multiple rolls of toilet paper. The science writer Nigel Calder tried comparing the passage of a billion years to a stroll down the island of Manhattan. To your right, ladies and gentlemen, you'll see the George Washington Bridge, and the first signs of unicellular life forms! Hiking past Central Park, Times Square, the Empire State Building: and more unicellular life forms! Other chroniclers have condensed the history of Earth into a single year, while still others have compacted it into a single day.

My favorite time-warp device is the one conceived by Kip Hodges, of imagining Earth as a human being with a seventy-five-year life span. "It's a real eye opener to think of the pace of our planet's development and the pace of evolution, in human terms," he said. By this reckoning, where twelve months is the equivalent of 60 million years, Baby Earth fattened up on a very fast track. It had finished condensing from the planetary disk around the sun and accreting added bits of rocks and metals to reach its present dimensions by one year of age. A month or two later, our big burbling bundle had belched up from its bowels a thick atmosphere of carbon dioxide, steam, nitrogen, sulfur, methane, and a smattering of other elements, a miasmic mix that our lungs would find utterly unacceptable but that allowed liquid water to wallow in the craterous basins on its surface rather than boil away into space. Early in its adolescence, Earth did what a human teenager should not do, and, somewhere, somehow, its saturated, still febrile tissue gave birth to the earliest forms of life. Roughly eight to ten weeks postpartum, blue-green strains of bacteria began spitting oxygen into the atmosphere, sparking a biochemical revolution that life eventually would put to spectacular use. Not until age sixty-three, however — about 700 million years ago — would we see the debut of multicellular animals. Mother Earth reached a grandmotherly seventy-two years before dinosaurs appeared, and the first ape didn't arrive until May or June of the final year, age seventy-five, of our handily, anthropocentrically foreshortened Life of Gaia. Modern *Homo sapiens* awaited the chiming in of December 31, agriculture and animal husbandry arose at 10:00 P.M. that night, the first writing was scrawled and the first wheel turned an hour later, the American Revolution was fought at 11:58 P.M., and Neil Armstrong muddied up the moon and muddled his way into Bartlett's at twenty seconds to midnight.

Looked at from this perspective, it wasn't just Rome that was built in a day, but all of human history.

As ancient as Earth may be, the universe, of course, is more ancient still. But it is not outrageously more ancient. It is not an order of magnitude, or ten times, more ancient than is Earth. Instead, it is only some three times older: 13.7 billion years have passed since the Big Bang gave leave for all being to begin. Personally, I've never been impressed with the age of the universe. To the contrary, its youthfulness makes me uncomfortable, the way I feel when I see the captain entering the cockpit of the plane I'm boarding, and he looks just old enough to no longer need a car seat. Only 13.7 billion years have passed since the beginning of everything — all time, all laws, all complaints? Yet when I've asked astronomers whether they agreed that the universe is remarkably underage for something so universal, they've stared at me as though it were a trick question, or a tedious exercise in metaphysics, before answering, Well, no, now that you mention it, it doesn't seem especially young to me at all. And the reason they see 1.37×10^{10} as being a perfectly reasonable vintage is that when it comes to cosmology, if it's about time, it's about space, and the spaces that the stuff of the universe has managed to cast itself across in these peri–14 billion years are so very, very great.

For astronomers, it is difficult to do justice to the scales of cosmic distances. Nearly everything is far away, farther away than you think no matter how innate your anomie. The one exception to this fearful farness is the moon. Our moon is only 240,000 miles away, or ten times the circumference of Earth; if you could fly there by ordinary jet, it would take twenty days. But that's it for the whimsical honeymoon options, practically speaking. A journey by jet to the sun would last twenty-one years, at which point passengers should be advised that contents in the overhead compartment, and the compartment itself, may have melted.

To gain a richer sense of cosmic proportions, we can paraphrase William Blake, and see the Earth as a fine grain of sand. The sun, then, would be an orange-sized object twenty feet away, while Jupiter, the biggest planet of the solar system, would be a pebble eighty-four feet in the other direction — almost the length of a basketball court — and the outermost orbs of the solar system, Neptune and Pluto, would be larger and smaller grains, respectively, found at a distance of two and a quarter blocks from Granule Earth. Beyond that, the gaps between scenic vistas become absurd, and it's best to settle in for a nice, comfy coma. Assuming our little orrery of a solar system is tucked into a quiet neighborhood in Newark, New Jersey, you won't reach the next stars — the Alpha Centauri triple star system — until somewhere just west of Omaha, or the star after that until the foothills of the Rockies. And in between astronomical objects is lots and lots of space, silky, sullen, inky-dinky space, plenty of nothing, nulls within voids. Just as the dominion of the very small, the interior of the atom, is composed almost entirely of empty space, so, too, is the kingdom of the heavens. Nature, it seems, adores a vacuum.

"The universe is a pretty empty place, and that's something most people don't get," said Michael Brown of Caltech. "You go watch *Star Wars,* and you see the heroes flying through an asteroid belt, and they're twisting and turning nonstop to avoid colliding with asteroids." In reality, he said, when the *Galileo* spacecraft flew through our solar system's asteroid belt in the early 1990s, NASA spent millions of dollars in a manic effort to steer the ship close enough to one of the rubble rocks to take photos and maybe sample a bit of its dust. "And when they got lucky and the spacecraft actually passed by *two* asteroids, it was considered truly amazing," said Brown. "For most of *Galileo*'s journey, there was nothing. Nothing to see, nothing to take pretty pictures of. And we're talking about the solar system, which is a fairly dense region of the universe."

Don't be fooled by the gorgeous pictures of dazzling pinwheel galaxies with sunnyside bulges in their midsections, either. They, too, are mostly ghostly: the average separation between stars is about 100,000 times greater than the distance between us and the sun. Yes, our Milky Way has about 300 billion stars to its credit, but those stars are dispersed across a chasmic piece of property 100,000 light-years in diameter. That's roughly 6 trillion miles (the distance light travels in a year) multiplied by 100,000, or 6×10^{17}, miles wide. Even using the shrunken scale of a citrus sun lying just twenty feet away from our sand-grain Earth, crossing the galaxy would require a trip of more than 24 million miles.

Interestingly, the distances between galaxies are relatively manageable compared to the gulfs between stars within a galaxy. That is, the average distance from one galaxy to the next is only a few tens of times larger than is the size of either galaxy, while the separation between stars is hundreds of thousands or millions of times greater than any single stellar diameter. "This is why stars don't run into each other, but galaxies do," said Robert Mathieu, a professor of astronomy at the University of Wisconsin. Our own Milky Way is expected to collide someday with its nearest neighbor, M31 — more familiarly known as the Andromeda galaxy — but we're talking about an awfully delayed train wreck, maybe 4 billion years in the future. Moreover, given their individual porousness, the fact that there are such wide breaches between each galaxy's solar wares, it will not be a particularly violent event, either.

In large part, cosmological metrics remain tragic, pious, almost impossible to forgive. With an estimated 100 billion galaxies in the universe, each outfitted with some 100 billion to 200 billion stars, we have a stellar inventory of 10^{22} far-flung suns: so many stars to yearn toward, so many ways to get lost in the dark. The distances from one to the next are so forbidding that, even if the universe teems with intelligent life, we are less likely to hear from an alien civilization than parents are from their college-age children. Yet before we curl into a Beckettian state of amniotic gloom, we might consider the perspective of Maarten Schmidt, one of the grandmasters of astrophysics. Schmidt has argued that, far from being a vast moor of anesthetizing proportions, the universe is unexpectedly compact, even homey. Schmidt, a courtly Dutchman in his seventies, has white hair and smock-blue eyes, and as he sat and talked in a calm, quietly animated voice, he kept his long arms folded neatly over his long, crossed legs. If you go outside at night, he said, and the sky is clear and you're far from a city, you can see Andromeda, that next-door galaxy we're expected to bump into someday.

"To get from Andromeda to what we would call the edge of the observable universe, you need only go out by a factor of three thousand," he explained. "The edge of the known universe, then, the most distant point from which light has been able to reach us, is only three thousand times as far from us as is our closest galaxy.

"Now suppose you look at the house nearest yours, and suppose it's a hundred yards away. If you were to go out three thousand times that distance, you'd have traveled only three hundred thousand yards, or about two hundred miles. So if you drew a circle around your entire community, your entire world, and that circle had a diameter of two hundred miles, wouldn't you think your community quite manageably proportioned? Wouldn't you be surprised at how close the edge of your world turned out to be? This is why I've argued that our universe, at least what we can see of it, is small.

"Of course, I realize that my position is totally indefensible," he said, before promptly defending it with a small, courtly smile.

Appealing though Schmidt's case may be in comparing the cosmos to a kind of picket-fenced pueblo, when it comes to professional-grade smallness, the bona fide articles are molecules and particles. You think you're living a normal, life-sized life, on human terms, driving to the supermarket and foraging for nuts, tubers, and pork chops; but in fact "life-sized" has nothing to do with you, the contents of your shopping cart, or Charlemagne's feet. The real merchants of life, the objects that keep life alive and qualify as life-sized, are all invisible. They are too tiny to be viewed with the naked eye, are instead microscopic, which means, of course, you need a microscope to see them. Unfortunately, for most of us, invisible often translates into insignificant — or, as my grandmother so musically put it, "Feh." Thus we are left with scant sense of just how invisible the components of which we are constructed really are. How big is a cell, or a protein sticking out of the fatty surface of said cell, or the DNA molecule at the center of the cell? When you look at the tip of your finger, roughly how many skin cells are you seeing? How about a bacterium — bigger or smaller than one of those rough skin cells? The water molecules that bond, de-bond, and re-bond so rapidly: Where do they fit on the mise en scène of the unseen?

To orient ourselves in the prefecture of feh, let's exploit the old theological dance floor, the head of a pin. A pinhead is two millimeters, or two-thousandths of a meter, across. By comparison, the average human hair is one hundred microns wide (a micron, you'll recall, being one-millionth of a meter). You could, then, drape twenty hairs over a pinhead if you pack them close together. Half the diameter of a human hair, or fifty microns, represents pretty much the lower limit of even the sharpest human eye's natural resolving power; anything smaller, and, well, there's a reason why "the width of a human hair" is so often used to mean "extremely narrow by naked-eye standards." In other words, you can't see an individual speck of ragweed pollen, which is twenty microns wide, without some sort of magnifying device. But among the allergy-prone, you needn't be seeing to be sneezing, and the 10,000 or so pollen grains that might cling to your pinhead are quite enough for "Gesundheit."

A human white blood cell is twelve microns wide. If the surface of your pinhead were wallpapered with white blood cells, you would be looking at about 28,000 of them. The wiener-shaped *E. coli* bacterial cell is two microns long by half a micron across, allowing 3 million of them to colonize your sewing notion; and, given *E. coli*'s pervasiveness, they've probably done exactly that. Bacteria as a rule are much bigger than the other microscopic characters that we designate as "germs" — the viruses. Unlike a bacterium, a virus is not a cell. It lacks nearly all the ingredients of a cell, most notably any means of autonomous replication, and instead must infiltrate the cells of other organisms and hijack the resident reproductive machinery for the sake of personal perpetuity. Shiftiness demands thriftiness: even a large virus like the Ebola pathogen has just one-tenth the footprint of *E. coli.* A tiny virus like the rhinovirus that causes the uncommonly contagious common cold is only three-hundredths of a micron, or thirty nanometers across, and tens of millions of them can sail through the air on a droplet sneezed forth by your ruby-nosed coworker.

Cracking open a human cell, you'll find the labor force of life, the heroic biomolecules that do all the work of keeping you alive for the 3 billion seconds of your life, give or take a few 10^{25} attoseconds. Hemoglobin, the blood-borne protein that captures oxygen molecules from the lungs and delivers them throughout the body, measures about five nanometers in diameter, a sixth the size of a cold virus. Collagen, the connective protein that gives both skin and Jell-O their bounce, is long, thin, and tough, like a piece of superfloss a few nanometers across and hundreds of nanometers long.

Deep within the belly of virtually every cell is our DNA — the celebrated, if symbolically overperfumed, corkscrew of a molecule that holds all our genes. This double helix is squeezed into a knobbly bundle that, depending on what the cell is doing, may measure anywhere from 100 to 1,000 nanometers in diameter. Even toward the upper range of DNA packaging, maybe 5 million of these little human genomes — 5 million Holy Grails, 5 million Books of Life, 5 million blueprints for a baby — could perch on a pinhead.

Proteins and DNA are blubbering baleens compared to other molecules of the cell. A glucose molecule, the simple sugar that fuels activity within your ever busy body cells, is only one-sixth the size of a hemoglobin protein, and the oxygen molecule that hemoglobin carries is one-third the size of that sugar.

Oxygen molecules are our clearest, shortest links to life. Deprive any part of our bodies of oxygen, and the asphyxiated tissue will start to die within minutes. What are these indispensable links but cufflinks, each molecule a dapper doublet of oxygen atoms, of O_2s, linking us to life, and linking us, on our scalar sally, downward to atoms. Sirs and madams, we're all made of atoms, and atoms are *tiny*. But how silly to try conveying information by choosing the smallest and most illegible typeface possible, and me not even a writer of pharmaceutical inserts. Atoms are way beyond the font. There are more than one hundred different types of atoms, from lightweights like hydrogen and helium through welterweights like tin and iodine and out to such mumbling mooseheads as ununpentium and ununquadium, but they're all pretty much the same nearly nil size. You can fit more than three atoms in a nanometer, meaning it would take 10^{13}, or 10 trillion of them, to coat the disk of our pinhead. And the funny thing about an atom is that its outlandish smallness is still too big for it: almost all of its subnanometer span is taken up by empty space. The real meat of an atom is its core, its nucleus, which accounts for better than 99.9 percent of an atom's matter. When you step on your bathroom scale, you are essentially weighing the sum of your atomic nuclei. If you could strip them all from your body, go on a total denuclear diet, you'd be down to about 20 grams, the weight of four nickels. Or roughly the weight of the doornail that you would be as dead as.

Those remaining twenty grams belong to your electrons, the fundamental particles that orbit an atom's nucleus. An electron has less than $\frac{1}{1,800}$ the mass of a simple atomic nucleus. Yet the cloud of one or more of these fairy-Ariel electrons that surrounds the atomic core defines the edge of the atom and hence its size. And, oh, how vast is the gulf between chunky core and orbiting cloud. The diameter of the atomic nucleus is just $\frac{1}{100,000}$ the size of the entire electron-limned subnanometer atom. Viewed from the more impressive angle of volumetrics, we see that, while the nucleus may make up nearly all of an atom's mass, of the meaty matter we weigh and inveigh against, it takes up only a trillionth of its volume.

Here it is worth a final reversion to metaphor. If the nucleus of an atom were a basketball located at the center of Earth, the electrons would be cherry pits whizzing about in the outermost layer of Earth's atmosphere. Between our nuclear Wilson and the flying pits, however, there would be no Earth: no iron, nickel, magma, soil, sea, or sky. Once again, there would be nothing, literally, to speak of. Inner space, outer space, galactically, atomically, no matter. We live in a universe that is largely devoid of matter. Yet still the Milky Way glows, and still our hemoglobin flows, and when we hug our friends, our fingers don't sink into the vacuum with which all atoms are filled. If in touching their skin we are touching the void, why does it feel so complete?

4

Physics

And Nothing's Plenty for Me

LET'S SAY THAT an asteroid portentously resembling a *Tyrannosaurus rex,* a giant trilobite, or Steven Spielberg were to slam into Earth tomorrow, annihilating the bulk of human civilization and the billions of civilians therein. What small sliver of human culture would be most worth preserving? What single piece of knowledge, what insight into the nature of the universe, would prove most useful to the few survivors as they struggled to rebuild all hope and opus of *Homo sapiens*? Lovers of the arts might suggest the collected works of William Shakespeare or Johann Sebastian Bach. The medically minded might vote for antibiotics, anesthesia, a general recognition of what not to do with the contents of one's chamber pot. Richard Feynman, the great physicist and titularly designated Genius and Joker, took seriously the problem of post-apocalypse reconstruction. "If, in some cataclysm, all of scientific knowledge were to be destroyed, and only one sentence passed on to the next generation of creatures, what statement would contain the most information in the fewest words?" Feynman asked rhetorically during one of his famed lectures. "I believe it is the atomic hypothesis, or the atomic fact, or whatever you wish to call it, that all things are made of atoms. Little particles that move around in perpetual motion, attracting each other when a little distance apart but repelling upon being squeezed into one another." Take that one sentence, he said, stir in "just a little imagination and thinking," and you have *The History of Physics,* Phoenix Rising edition.

Physics is one of those modest disciplines that, in the words of a popular text by Steven Pollock, a physics professor at the University of Colorado, is nothing less than "the study of what the world is made of, how it works, and why things in the world behave the way they do." And the less said, the better. Physics revels in reductionism, a word that to many implies "simplistic" and "probably not applicable to anybody in my social circle," but really is another way of saying "understanding something complex in terms of its constituent parts." That, of course, is what most sciences seek to do, but physics goes the furthest, breaking apart the constituent parts until they're crying for Muster Mark.* Physics is the science of starter parts and basic forces, and thus it holds the answers to many basic questions. Why is the sky blue? Why do you get a shock when you trudge across a carpeted room and touch a metal doorknob? Why does a white T-shirt keep you cooler in the sun than a black one, even though the black one is so much more slimming?

As the science of starter parts and forces, physics can also be defended as the ideal starter science. Yet standard American pedagogy has long ruled otherwise. In most high schools, students begin with biology in tenth grade, follow it with chemistry, and cap it off in their senior year with physics, a trajectory determined by the traditional belief that young minds must be ushered gently from the "easiest" to the "hardest" science. More recently, though, many scientists have been campaigning for a flip in the educational sequence, teaching physics first, the life sciences last. Leading the charge for change is Leon Lederman, a Nobel laureate in physics and professor emeritus at the University of Illinois, who has the distinctive shock of almost fluorescent white hair with which the elder statesmen of physics are so often blessed.

Lederman and others argue that physics is the foundation on which chemistry and biology are built, and that it makes no sense to start slapping the walls together and hammering on the roof before you've poured the concrete base. They also insist that, taught right, physics is no "harder" than any other subject worth knowing. Some schools have adopted the recommended course correction, and others are sure to follow. I not only agree with the logic of Lederman's from-the-ground-up approach; I also trust his populist heart. Lederman, it so happens, has long been lobbying the networks to do their bit for science's public image by starting a television series based on a team of laboratory sci-

* In the mid-twentieth century, Murray Gell-Mann, a theoretical physicist and puckish promoter of *Finnegans Wake,* famously named the fundamental building blocks of matter "quarks," after a line from James Joyce's least readable novel: "Three quarks for Muster Mark! / Sure he hasn't got much of a bark / And sure any he has it's all beside the mark." Despite Joyce's presumed intent that "quark" be pronounced to rhyme with *Mark* and *bark,* the subatomic particle generally is articulated *kwôrk* (as in *pork*), which also happens to be the preferred pronunciation for the creamy, acid-cured cheese product popular in Germany.

entists. Physicists, biochemists, drama or sitcom, Lederman doesn't care; what counts is that the characters defy geek stereotypes with their emotional struggles and interpersonal parries, their drive and self-doubt, their prominent cheekbones and stylish footwear.

Physics, then, is the pylon science, the discipline on which the others are piled, if sometimes peevishly. And as Feynman proposed in his Duck and Recovery plan, the most fundamental facet of this foundational field is the atom.

Everything, every single thing deserving of the designation "thing," is made of atoms. Even those things that are not obviously thingly can, in the end, be stripped to their atomic Skivvies. Thoughts, for example. As they drift from your brain and through the Sheetrock of your office cubicle, they seem defiantly fleeting, robustly substance-free. Yet the brain cells that gave rise to your thoughts are all built of atoms, and if one thought triggers another it does so via the transmission of neurochemicals along synaptic pathways in your brain, which again are vast assemblages of atoms; and if you tap your thoughts down in an electronic journal for later dissemination as friendly spam, you despoil an innocent screen by rearranging the atoms of its phosphor-coated surface.

The atomic Tinkertoy set of which we are constructed happens to be a magnificent system for getting things right.

"If you want to replicate something, you will make fewer mistakes if it is made up of discrete units than if it is made up of continuous material," said Ramamurti Shankar, a professor of physics at Princeton University. "By analogy, you will make fewer mistakes if you are trying to spell a word than if you are trying to reproduce a color." It's good to know, at a gut level, he added, "that there are only a hundred-odd different letters, different types of atoms, to worry about."

That all matter is built of atoms is one of those profound insights into the nature of reality that gestated in larval, largely figmental form for some two thousand years, before twentieth-century physicists like Albert Einstein and Niels Bohr finally offered experimental evidence of the atom's existence. The Greek philosopher Democritus argued circa 400 B.C. that everything was made of invisible, indivisible particles, which varied in shape, size, and position and which could be mixed and matched to yield every manner of matter. Democritus called these particles *atomos*, meaning "unbreakable" or "uncuttable." Among the fiercest opponents of this early version of atomic theory was Aristotle, who, for all his brilliance, had a habit of dismissing some really fine ideas. Aristotle insisted that the world was composed, not of discrete particles, but of four essences or qualities — earth, fire, air, and water. Aristotle's woolly, wrong-headed, but admittedly evocative schema held sway for hundreds of years, and still claims a sizable fan base among followers of astrology.

The early models of atoms resembled our solar system, with the nucleus as the sun and the electrons orbiting like planets around it. Another familiar portrayal of the atom is the Spirograph-style icon from the 1950s, of a central disk surrounded by three or four ellipses, like the official emblem for Arco, Idaho, which proudly describes itself as the "First City in the World to Be Lit by Atomic Power." An atom doesn't look anything like a solar system or a kitschy city logo, though. You can't really say what it looks like, in the ordinary visuospatial sense of the phrase. Not merely because the atom is invisible to the unaided eye. Cells and bacteria are "invisible," too, but you can see a cell or a microbe perfectly well with the right microscope. The problem with atoms, as Brian Greene made plain to me, is that they are so small they fall into the perilous domain ruled by Werner Heisenberg's uncertainty principle: you view it, you skew it.

"If we could blow up an atom to something the size of, say, a paperweight on your coffee table," I asked Dr. Greene, even as I noted that his coffee table was free of any papers in need of weighting, "what would we see?"

"'See'?" he echoed, so slowly the word sounded multisyllabic. "What would we see? I don't want to sound Clintonesque here, but it depends on your definition of see.

"When we talk about seeing things in the everyday world, we're talking about light," he explained. "We're talking about photons of light, particles of light, banging into our eyes and allowing us to see. But when you get down to the scale of the atom, those photons can change the nature of the thing you're seeing." The electrons that surround the atom can absorb and emit photons, he said, and when they do, the electrons jump around, altering the atom's conformation. "We long to impose the everyday experience of sight on the tiny little atom, but to do so requires that we change the atom itself. We can't literally see down there."

OK, forget the literal paperweight, I said. What might we figuratively not really see?

"A cloud," he said. "A picture of an electron cloud is a reasonably accurate way to think about it." Like a tumbling dust bunny that you can never quite catch with your DustBuster? I asked. Or the cloudy smear on a television news report when the identity of a moving figure must be concealed? Well, sort of, Greene replied. But not like a swarm of gnats. Not an aggregate of many distinct objects. The image of an electron cloud is really a device to depict probability distributions, he said, telling you where an atom's electrons are likely to be found, and to give you a sense of how the electron's potential positions are distributed.

Even for the simplest atom, hydrogen, which has just one electron whizzing about the single proton of its nucleus, the electron has so many points it may be found, so many places it has been and will be again, that the entire boundary of the hydrogen atom can be envisioned as a spoonful of cloud.

Yet before we get carried away with this lovely image of electron distribution plot as Pre-Raphaelite hairdo, we must keep in mind that Aristotle erred: matter is not a sweep of qualities all blending seamlessly together. Atoms may and often do attract each other. Atoms form bonds, usually by sharing electrons consigned to each participant atom's outermost orbit. Through the artful bartering of electrons along their frontiers, two hydrogen atoms and one oxygen atom conjoin to form a molecule of water. But, importantly, the atoms do not merge, or invade one another's comparatively vast expanse of empty inner space. The atoms remain discrete entities, distinct particles composed of protons and neutrons in the nucleus, a huge amount of hollow space, and a cloud cover of electrons located far, far from the nucleus. The hollow space is, as a rule, a sacred place. Neither the electron clouds nor the nuclear particles of one atom will penetrate the inner void of another atom and take a tour, maybe sidle up to the foreign nucleus, wave hi and bye, and then head home again. Only under extraordinary conditions, as in the high-pressure furnace of the interior of a star, can two atoms be squashed together, at which reaction their nuclei combine to form a new, heavier type of atom, another element further along the periodic scale, a subject we'll visit later.

Most of the time, though, atoms maintain their autonomy and ethnic identity, including when they are in a stable molecular relationship with other atoms. The hydrogen and oxygen atoms with which the oceans are filled remain hydrogen and oxygen to their core and can be plucked free from one another, although it takes energy to cleave the bonds of a water molecule or any other molecule and isolate the constituents. It is an astonishing thought that every last backdrop and foreprop of our lives, the sweet air we breathe, the cool water we drink, the speed bumps we bump over, all consist of discrete, hollow particles, trillions upon quintillions of vacuum-filled atoms that will get close to each other, but never too close. As Feynman said, atoms will attract if they're a little distance apart, but if you start getting pushy, they push right back.

What is it that keeps atoms discrete, and in such anally compulsive need of their "space"? And why, if most of matter is empty, am I sitting here on a reasonably comfortable mahogany chair rather than falling plunkity-plunk through the hollow atoms of the furniture, floor, planet, to join poor Commander Frank Poole in his death drift across the velvet void of outer space?

The answer lies in the dispositional humors of the subatomic particles — the pieces of which an atom is constructed — and the ploys, counterploys, and compromises in which they tirelessly engage. In the nucleus are the heavyweights, the protons and neutrons, which manage to make up more than 99.9 percent of an atom's mass while occupying only a trillionth of its volume. All atoms are not the same, of course. Our world trembles and gleams with atoms of gold, silver, bismuth, platinum, lead, sodium, mercury, indium, iridium, xenon, carbon, silicon, and some one hundred other basic syllables of being that we call the elements. The elements are substances that refuse to be reduced to simpler substances through normal chemical or mechanical means. If you have a sample of pure lead, you can break it apart or melt it down into smaller lumps of lead, but each piece will still be composed of lead atoms, and not the gold you might covet or the strontium you probably don't, unless you're in the pyrotechnics business and appreciate its flammability. And while the different atoms are all about the same size — a tenth of a billionth of a meter across — they diverge in their mass, in the number of protons and neutrons with which their nucleus is crammed. Hydrogen, the lightest and by far the most common element in the universe, has the maximum minimalist of a nucleus, composed of a single proton; however, there are variants of the atom, given the unflattering designation of "heavy hydrogen," which possess one or even two neutrons in addition to the single proton. Many of the more familiar elements have pretty much the same number of protons and neutrons in their hub: carbon the egg carton, with six of one, half dozen of the other; nitrogen like a 1960s cocktail, Seven and Seven; oxygen an aria of paired octaves of protons and neutrons.

Yet these elements, too, can be found in fattened versions of themselves, called isotopes, the added bulk compliments of extra neutrons. Carbon, for example, exists in an eight-neutron isotope that is so unstable, and so predictably prone to jettisoning its eighth neutron, that archaeologists and paleontologists use its pace of neutron ejection, or decay, as a kind of clock to help them date ancient buried treasures, be they the caried teeth of a prehistoric king who had an obvious taste for sweets, or animal bones carved into the first dental tools, or the charred remains of the first dentist.

Among the nuclei of heavier elements, protons are usually outnumbered by neutrons, sometimes substantially so. Mercury's 80 protons, for example, are raised 1.5 to 1 by the slippery metal's 120 neutrons. But protons more than compensate for their minority status by their unshakable sense of self-worth. For while any type of atom may have a few neutrons more or less without losing its essential identity, its proton census is nonnegotiable, the most elemental element of an element. Proton content alone distinguishes one species of atom, one element, from another and therefore serves as the element's official atomic number. Gold holds 79 protons in its nucleus, and hence is given the atomic number of 79; while at slot 78 we find platinum, which, for all its 117 neutrons, is one proton shy of being, figuratively speaking, as good as gold.

So what grants protons their privileged status? If protons and neutrons are similarly proportioned, and equivalently responsible for the roughage in your broccoli floret or the buoyancy in your daughter's balloon, why is it proton tally alone that separates selenium, atomic number 34, an essential dietary nutrient that helps convert fats and protein into energy, from arsenic, atomic number 33, a highly toxic substance that is used to kill rats, weeds, and the occasional Roman emperor?

The answer is electric charge: a proton has it, while a neutron does not. The neutron, true to its name, is an electrically neutral particle, and if a neutron were to order a drink at the bar, as another MIT joke has it, and then ask how much it owed, the bartender would reply, "For you, no charge." The proton, true to a whimsical convention that dates back to Benjamin Franklin and his kite, is said to be a positively charged particle, while the atom's other electrically charged particle, the electron, is said to be negatively charged. The terms positive and negative are not judgment calls — a reflection of physicists' preference for one particle over another, or of the proton's capacity to improve property values while the electron leaves old car parts strewn on the lawn. The vocabulary could as easily have been reversed, and the proton designated as negatively charged and the electron pronounced positive, but they weren't, so let's not. What is important is that the charge of one counterbalances the charge of the other. An electron may be more than one thousand times lighter than a proton, but its charge is every bit of a match for that of the nuclear giant. And well matched they are, for protons and electrons attract each other, just as opposites are legendarily said to do in the macroscopic community, although in that case the reaction all too often ends up requiring the intervention of other macroscopic units known as divorce attorneys.

But what exactly is this subatomic charge, this positive charge of the proton that attracts and tit-for-tats the negative charge of the electron? When you talk about a fully charged battery, you probably have in mind a battery loaded with a stored source of energy that you can slip into the compartment of your digital camera to take many exciting closeups of flowers. In saying that the proton and electron are charged particles while the neutron is not, however, doesn't mean that the proton and electron are little batteries of energy compared to the neutron. A particle's charge is not a measure of the particle's energy content. Instead, the definition of charge is almost circular. A particle is deemed charged by its capacity to attract or repel other charged particles. "A charge is an attitude; it is not in itself anything," said Ramamurti Shankar. "It's like saying a person has charisma."

Another way of defining charisma is "force of personality," which brings us to the reason why charged particles react to other charged particles. They are obeying the laws of electromagnetism, one of the four fundamental "forces" of nature. You've probably heard about these four fundamental forces, and you might know them by name: electromagnetism, gravity, the strong force, and the weak force. But "force," like "charge," is one of those words that comes up so often in everyday conversation that its meaning seems deceptively self-evident, and it is rarely explained in the context of fundamentalism. What distinguishes a fundamental force of nature from the more familiar, frightening forces of nature, like hurricanes, earthquakes, Donald Trump's hairpiece?

A fundamental force is best thought of as a fundamental interaction, a relationship between two chunks of matter. It turns out that there are just four known ways that one piece of matter can communicate with another, four approaches to acknowledging the existence of a body other than one's own. Each of these interactions differs in strength and range, operates according to a distinct set of rules, and yields distinct results. They are not, however, mutually exclusive. For example, all bodies, no matter how minute, are gravitationally attracted to one another. Charged or neutral, spinning or sedate, masses make passes through the universal come-on of gravity. Yet difficult though it may be to believe if you've ever tried putting on a fake set of wings and then flapping your arms while jumping from the roof of your house, gravity is by far the weakest of the four fundamental forces. It makes its impact felt only on relatively large chunks of matter like stars, planets, and chowderheads leaping from rooftops.

If you take a couple of electrically charged particles, on the other hand, they're gravitationally attracted to each other, sure. But, being charged, they're also under the sway of the electromagnetic force, which is, oh, about 10^{40}, or more than a trillion trillion trillion, times stronger than gravity. Depending on whether the particles are of the opposite or similar charge, the electromagnetic force will either pull them closer or push them apart, gravity be damned.

Scale and context always dictate which force will be with you. In one sense, for example, the strong force merits its swaggering codpiece of a name, for it is the strongest binder known in the universe, more than a hundred times stronger than electromagnetism. The rules of its engagement keep protons and neutrons glued together in the nucleus, overriding the electromagnetic repulsion that otherwise would send all those positively charged protons fleeing, one from the other. But the strong force operates only across the ludicrously short distances between and within the particles of the nucleus. As for the weak force, the prompter of neutron decay and the fussy obscurantist of the force quartet, its reach is also limited to nuclear dimensions.

Physicists propose that the four forces are really four manifestations of a single underlying superforce, and that when our universe was young, firm, and hot, the forces behaved as one, too; only with the inevitable aging and cooling and spreading of the cosmos did the single force fracture into four separate instruments. Scientists' quest to unite the four forces into a single equation, a Grand Unifying Theory succinct enough to fit onto one of those scratchy XXL Beefy T-shirts with the too high collar that nobody wants to wear, is an effort to discover the primal commonality underlying the current plurality.

Whether they succeed or not in tracing the math to glory, we live in a world of four fundamental forces, four distinct means of matter-to-matter communication; and whatever parleying occurs among particles, and the organisms constructed of those particles, occurs through one or more of the four. The ball you threw into the air, as it makes its way up and down, is responding to the lure of gravity. But what of the force that sent the ball flying in the first place? You the pitcher applied a force to it in the classic, Newtonian sense of the word, meaning you flexed your muscle and caused a stationary object to start moving. Through what fundamental force, though, did the particles of your hand convey their message to those of the ball? You may be Ty Cobb, Pete Rose, or the decidedly Beefy-T David Wells, but sorry, it's not the strong force.

For the source of our sporty fling, and of the many other ways we seize the day and size it up with all five senses, we must look again to the atom's architecture, and the loves, snubs, and limits that keep it standing.

The electron, with a designated minus sign tattooed on its forehead, finds the positive proton terribly attractive, and wants to spend its time somewhere in the vicinity of one. Yet the electron is also in constant, twirling motion, and how grateful we should be for its vigor. You'd think that these oppositely charged particles would fall into each other's arms, that the electron, smitten by the Grace Kelly glow emanating from the nucleus, would simply dive toward the proton and not stop until it had reached its destination. You'd think that all atoms would collapse like popped bubble wrap, taking every one of us precious parcels down with them. But no, the tremendous momentum of the electron throws it into orbit around the nucleus, keeping it at a distance and on the fly, just as the angular momentum of the planets ensures that they continue wheeling around the sun to which they are gravitationally attracted, rather than plunging into its fiery depths like kernels into a corn-burning stove. Electrons can never stop to catch their breath. For one thing, they have no lungs. For another, if electrons did stop moving, you'd be able to tell both where the particles were and how fast they were — or rather, were not — going. Heisenberg said in no uncertain terms that you can't know both details about an electron simultaneously, so, oops, gotta dash. Electron speed changes depending on how excited the particles are: in the laboratory they can be propelled toward the velocity of light, but even on an ordinary day spent clouding around the atom, they race about at 1,370 miles per second — fast enough to circle Earth in 18 seconds.

Yet electron pace is hardly the sole determinant of an atom's configuration. The entire carousel flirtation between proton and electron is as vigorously supervised and ritualized as an antebellum courtship. Electrons cannot flit about wherever they please, but are confined to specific zones, or shells, around the protons to which they are so attracted. The shells are arrayed one inside the other, and each is able to accommodate a set number of electrons. The shell closest to the nucleus has room for just two electrons, the two subsequent beltways have space for eight negative particles apiece, while those farther out can manage eighteen or more electrons. Once a shell is filled, even the president and his cloud of armored SUVs couldn't nose their way in. An electron also cannot travel in between shells, just as you cannot stand between two steps of a staircase. An electron can, however, switch from one byway to another, assuming there's room. Sometimes, if an atom is blasted with a beam of light, a few of its electrons may become stimulated and jump to vacancies in shells farther from the nucleus. But "jump," in the quirky subatomic subculture at the base of all being, does not mean "bound in a continuous motion from here to there"; it means "disappear momentarily from the shell I was in and reappear suddenly in the shell above me." This Houdini maneuver is the famed "quantum leap," for the electron is shifting from one permissible shell, or energy level, or quantum, to the next, without trying to ram through concrete barriers between lanes. The expression "quantum leap" long ago found its way into popular language, usually to mean something like "a really big change" or "a great jump forward," and though some people have griped that it's a misuse of language because the distance between electron shells is so vanishingly small, I'd say the criticism is misplaced. Quantity notwithstanding, a genuine quantum leap is qualitatively spectacular, a bit of *Bewitched* without the insufferable husband.

An atom's demand for electrons, and thus the number of orbital shells that surround it, emanates from its protons. As it happens, an atom is like Switzerland; it prefers to assume a neutral stance whenever possible. This preference requires that each of its protons, the comparatively massive, electrically charged, and imperious components of the nucleus, be paired with an electron. An atom of gold, with its 79 protons, requires 79 electrons to reach its favored state of neutrality. An atom of gold, then, is a snaggle-toothed hundred millionth of a centimeter of a beast, comprising a nucleus of 79 protons and 118 neutrons, and then, far, far from the dense, thumping heart, 6 cloudy shells, 6 probability pathways along which 79 electrons spin.

Yet even with all its intricacy, its swirling bazaar of particles, an atom of gold, like all atoms, is hollow, is nearly nothing, is emptier than a fraternity beer keg on Sunday morning. Why, then, do the two gold rings on my fingers — one my wedding ring, the other a gift from my husband in honor of our daughter's birth — feel reassuringly firm and enduring, slim, smooth circles that I never remove, yet which are palpably, as well as symbolically, not-me? Sometimes, in winter, my fingers shrink enough that the rings slide about and threaten to slink right over the knuckle and down the sink. But neither ring ever evinces the slightest inclination to fall right through the diameter of my finger like a hot knife through butter, as hollow as my finger atoms and ring atoms may be. So what gives, or rather, what doesn't give?

The answer, again, is charge, this time of electrons. All atomic nuclei are surrounded by clouds of negatively charged electrons, and like charges repel. The electromagnetic force is second only to the strong force in its exertions, and so the repulsion is serious. "Electrons don't like being around other electrons," said Shankar. "Atoms keep a comfortable distance from each other because of their electrons. It is really the electromagnetic force that keeps you from falling through the floor."

As we'll see in the next chapter, chemistry explains why the atoms of our fingers or those of a piece of wood manage to stay together and maintain a semblance of solidity. Nevertheless, said Brian Greene, "If you could imagine zooming in close enough to see the atoms of your fingers interacting with the atoms in this table, or this chair" — he touched each item of furniture in turn — "you would see their outer electrons repelling each other with electromagnetic force. Any time you touch or feel something, that is the electromagnetic force at work.

"In fact, electromagnetism governs all our senses," he said. Sight: the electromagnetic waves that we call light waves convey their message by interacting with the electrons in the atoms of our retina. Hearing: atoms of air press against the atoms of our auditory canal, and the consequent skirmishing among electrons is interpreted by the brain as Bach's Sonata for oboe and harpsichord. Taste and smell: food atoms poke their electrons against the atoms of the taste buds on our tongue and of the olfactory receptors in our nose, and the specific pattern of taste and smell receptors thus chafed informs the brain, Roast chicken. Gee. I haven't had any of that since last night.

Even as protons and neutrons form the overwhelming bulk of matter, of our bodies, the floor beneath our feet, the stained upholstery on our seat, the leftovers we are about to eat, it is the electrons, those fidgety flecks that account for less than a tenth of 1 percent of an atom's mass, that allow us to perceive and embrace the world around us. To frame it another way, it is one electron's antipathy toward another that shields us from the essential emptiness of every entity, that allows the protons and neutrons to puff with pride and play the pooh-bah, the fat of the land. How are they to know that, when you look at a table or any other object, you're not seeing the regal nuclear particles, but rather light bouncing off the electron mask with which each atom is adorned.

Consider, then, that, while the force of gravity may keep our feet on the ground and our planet skating around the sun, electron hostility is what makes the trip worthwhile.

I hate winter, the whole surgical tool kit of it: the scalpel cold, the retractor wind, the trocar dankness. I hate the snow, whether it's fluffy virginal or doggy urinal. I hate the inevitable harangues about how you lose 30, 50, 200 percent of your body heat through your head, because above all I hate winter hats and refuse to wear one. What happens with a hat? You take it off, and half your hair leaps up and waves about like the cilia of a paramecium, while the rest lies flat against your skull as though laminated in place.

The last section ended with a paean to electron hostility. This one begins with a peeve about electron mobility, the source of what is paradoxically, and somewhat inaccurately, called static electricity. The peeve can't last long, though, for I love subatomic nomadism when it's not wasting time raising manes or sticking skirts to stockings but is instead making itself useful by toasting bagels and running blenders, or for that matter allowing brain cells to fire or muscle cells to extend or contract. And guess whose fleet feet are behind the flicking of a switch that we in the West so take for granted and depend on that major blackouts cost billions in lost business? The electron, from the Greek word for "amber," the fossilized trickles of tree sap that, according to Greek myth, were the sun-dried tears of the gods, and that, according to Greek experience, were readily charged when rubbed with cloth.

Electrons are extremely tiny. They have mass, but the amount is so modest that they can sometimes behave almost like photons, the massless particles that carry light. Moreover, as far as we know, electrons are elementary particles, meaning they can't be broken down into even smaller particles. Scientists can crack apart the nuclear particles, the protons and neutrons, into even smaller subnuclear particles, called quarks. But no matter how they have slammed and shazammed electrons in the brutal conditions of a high-energy particle accelerator, they have not found subelectronic components inside.

Electrons have internal integrity, but atomic fealty is another matter. For electrons, one proton is as good as the next, and though the attraction between the negatively and positively charged particles is reasonably strong, it is also, in some cases, a surprisingly easy connection to rupture. If you drag a comb through your dry hair, the comb will strip off millions of electrons from the outermost shells of the atoms of your coiffure. That comb is now bristling with extra electrons, and thus is a negatively charged device. If you then hold the comb close to a few snips of paper, the pieces will hesitate for a moment, and then jump up and stick to the teeth of the comb. This act of levitation is evidence of electronic itinerancy. During that initial hesitation, many electrons on the surface of the paper clippings were repelled by the bolus of electrons presented to them on the negatively charged comb, and jumped aside, either to the edges of the paper, or off the scraps altogether. As a result, the surface atoms of the paper pieces suddenly found themselves in a state of electron deficit; and how better to resolve the crisis than by bounding toward the surfeit of negative particles beckoning from the comb overhead? Yes, the very same grooming tool that had thrust the paper into a predicament of positivity in the first place. It's like entrepreneurial capitalism. Forget about finding an extant consumer demand to be filled; go out there and dig a new need from scratch.

The Winter Hat Trick is a slightly different composite of repulsive reactions and rebound attractions. Why do some hairs stick up and out when you remove your wool cap? In being pulled off your head, the wool scrapes electrons from the outer layers of your hair, transforming each strand into an electron-depleted, positively charged object. Positive repels positive just as negative repels negative, so the strands try to get as far from one another as possible. At the same time, the positively charged hairs closer to your scalp and face become unusually attractive to the electrons in your skin, drawing the strands in close enough so the electrons can hopscotch between flesh and tress.

Importantly, charged atoms seek to fill their vacant shells or to shed their excess electrons and return to the bliss of Swiss neutrality; and so, too, of necessity, do the objects to which the disgruntled particles are hitched. Some materials are more apt than others to help alleviate one's imbalance of electric charges, a facility that usually requires a high tolerance for the most totable of motes, electrons. The metal of a doorknob is an excellent Ellis Island for electrons, for when metal atoms arrange themselves into molecules, the electrons in their outer shells are often loosely attached and free to roam about from one metal atom to another. The neighborly sharing of electrons tends to strengthen the bonds among atoms, lending metals their legendary toughness and historic utility to the military profession. A metal's steady electron churning also means there are always holes to be found, regions of positive charge toward which immigrant streams of electrons will be drawn. Metals, in short, are superb conductors of electron flow.

Dry air is an abysmal conductor of electrons. The reason why static cling and shocking handshakes are a particular problem in winter is that indoor, heated air tends to be extremely dry. Thus, any charged particles you may have gathered on your person by walking across a carpeted room or removing an overcoat will likely remain on your person unless or until they have somewhere else to go. They'll tug remaining layers of clothes together, or they'll jump from you to the proffered hand of a newcomer — especially if that person is wearing a metal ring. In that single shocking moment, about a trillion electrons typically leap to their new host, bringing the donor back to the near neutrality that usually characterizes the human body.

On a muggy summer day, by contrast, water molecules in the air, which happen to be slightly positively charged on one end, tend to wash the odd electron from foot or fabric fairly quickly, returning you to a state of private tranquillity before you reach the door.

Not so for the world outside. The forked lightning of a fabulous summer thunderstorm offers the most spectacular example in nature of how a swelling buildup of charge disparities — between slip-sliding stacks of air masses, and between turbulent atmosphere above and stolid terrasphere below — is resolved by the abrupt conductance of negative and positive charges from cloud to ground and back up again, all courtesy of the bucket brigade of water droplets in the intervening sky. Lightning, in other words, is a doorknob spark on a very big scale. You can call it the vente grande version of static electricity, if you like. But you do so at your peril: The word "electricity," like "force," like "charge," means different things to different people, or to the same people in different moods and typefaces. And for some scientists and engineers, the constant, careless bandying of the term "electricity" makes them so ANGRY they could BLOW A FUSE. William J. Beaty, an electrical engineer who keeps an uppercase-heavy Web site devoted to explaining electromagnetism and the many myths and misconceptions surrounding it, complains that electricity is a "catchall word" applied helter-skelter to a welter of very different phenomena: electrical energy, electric current, electric charge, electric bills. For the sake of clarity, he argues, maybe we'd be better off if we all agreed that "ACTUALLY, 'ELECTRICITY' DOES NOT EXIST!"

Beaty is mostly right on two counts. First, the term is terribly vague, far more so than many other examples of crossover vocabulary. Whereas "force" and "charge" retain specific scientific meanings apart from their everyday sentiments, "electricity" does not. It survives among scientists mainly as a folksyism, the way herpetologists still talk to the public about "reptiles," even though they abandoned the term among themselves years ago as archaic and imprecise.

Second, as far as most people are concerned, the thing they are likeliest to call "electricity" barely seems to exist until it ACTUALLY does not EXIST! That is, when they are at their desk laboring frantically on their computer — too frantically to bother saving their work now and again — and suddenly, "Hey, what the Helmholtz happened to all the electricity??!!" For many of us, electricity is the invisible power we expect to purr forth bounteously from wall outlets or spring-loaded sets of alkaline batteries and light our lamps, warm our rooms, chill our food, clean our clothes, and run the little digital clocks embedded in at least seventy-four household devices, including the cat box. "Electricity" is what we try to guard our toddlers against by plugging up empty outlets with pronged pieces of plastic. If electricity isn't the right word, what is? Of greater relevance, what is this thing we've been misnaming all these years?

Here is where many researchers might question Beaty's uppercased rebuff of the word. As they see it, the topic of electricity and how it works offers a beautiful opportunity for conveying many fundamental principles of physics in a single jolt. Sure, "electricity" is used at the whimsical and often confused discretion of the speaker to describe a motley set of physical events, but if you take each of those effects in turn and reflect on where it fits into the ordinary extraordinary act of filling a room with a fistful of fabricated sun, you may surmount your conviction that you'll always be left in the dark.

I mentioned earlier that lightning is a kind of Wagnerian show of static electricity, although scientists like Beaty would howl, again rightly, that there is nothing "static" about it.

Nevertheless, people of varying degrees of electrical expertise have long distinguished between the flying sparks and thunderbolts that they can't control, and the electric currents that they generally can (unless they work for my utility company). An electric current, like static cling, arises from the peregrinations of charged particles, but in an electric current the flow of particles is continuous and targeted and expensive; static cling is episodic and unmanageable, a little offer-with-purchase that you can't quite refuse.

Comprehending and domesticating electricity for the betterment of the better-off segments of humankind took centuries of work by a procession of scientists whose names have been immortalized in an enviable format: as standard international units of measurement to be memorized by physics students on the eve of finals. There was Count Alessandro Volta, an Italian physicist who invented the chemical battery; James Prescott Joule, a British physicist who showed that heat is a form of energy; Charles Augustin de Coulomb, a French physicist and pioneer in the study of magnets and electrical repulsion; James Watt, a British engineer and inventor who designed and patented a very good steam engine but who never worked as secretary of the interior for the Reagan administration; André Marie Ampère, a French mathematician and physicist who discovered the relationship between magnetic force and electric current and whom we honor in the ampere, or amp, but not, curiously enough, in the verb "amplify," which derives from the Latin word for "enough"; Luigi Galvani, who discovered that electric currents prompt contractions in nerves and muscles, and whose name is behind the thesaurus-friendly "galvanize"; and Georg Simon Ohm, a German physicist who determined the relationship between voltage, current, and resistance in an electrical current, and who is rumored to have practiced yogic meditation when he thought nobody was around.

From Ohm we get ohm, the unit used to measure resistance in an electrical circuit or device. And though no one expects you to master the nuances of units or their namesakes (except to remember who the real watt's Watt was and what that Watt was not), the ohm is a good place to start talking about the electricity coursing through your cords, and what it says about all of us.

Resistance, in the broad, Newtonian sense, is a force, like friction, that works in the opposite direction of a moving body and tends to slow the body down. When it comes to an electric current, resistance is a measure of how much a material impedes the free flow of electrons from input to household device. The higher the resistance and the larger the ohm, the slower the flow. Dry air, as I've discussed, has extremely high resistance to electron passage. Metals tend to have low resistance and conduct electron flow readily. Some metals are better conductors than others because they have comparatively more gaps in their shells, which are open for electron trafficking. The shells of copper and tungsten, for example, are particularly well honeycombed, bearing vacancies in their penultimate as well as their outermost shells. So it is that copper is often spun into electrical wiring, while the comparatively rarer tungsten is reserved for the delicate filaments of a light bulb.

Of course, you don't want unregulated electrons cruising in your home — exposed bundles of wires that, if touched, will divert some of their sizzle in your direction. The metal conduits of a common cord come wrapped in layers of insulation, a material with a comfortably large ohm, such as rubber or plastic, where the atoms clutch their electrons tightly and have no desire or patience for the passage of wayfaring particles. Electrons may gather readily on the surface of a balloon, but the rubber resists their transdermal penetration — as does the rubber or similar insular material wrapped around an electric wire.

But what does it mean to talk about a flow of electrons, or an electric current? It means that energetic charged particles are being directed along a pathway, such as a length of wire, usually toward a target, where they will be expected to do some work. Not all or even most of these excited electrons will traverse the entire length of the conducting corridor. Some may flow all the way through; the majority will jostle an atom en route, loosening an orbiting electron and pushing it forward, then moving into the vacancy itself. The important point here is that a large stream of trillions of electrons is either getting a push from behind or a pull from up ahead, and is excited, anxious, driven.

Now, electrons are always on the move, no matter what. As they cloud around an atom, they refuse to stop. The electrons in a discarded fragment of copper wire are jiggling about continuously, hopping from one metallic atom to its neighbor, rearing back at any sign of other electrons, with the territorial indignation of cats. Yet these are ordinary electron motions, powerful enough to fulfill the edicts of their atoms, but not enough to do more, not enough to flip a switch or turn a gear. If the electrons are to take on any organized, extracurricular activities, they must feel inspired. They must be animated. They must eat.

Electrons have mass — an exceedingly small amount of mass, but mass nonetheless. Electrons, then, are a form of matter. They are condensed fragments of the cosmos that need some reason to get out of bed in the morning and off the couch in the evening. They are not self-motivated. That is to say, they are not a form of energy, or at least not of useful energy.

The vast splintered vale of our universe, as far as we know, is stocked with two basic offerings, two categorical insults to His Lowly Holiness of Absolute Nothingness that might otherwise have held sway: matter and energy. For all the comparative emptiness out there and in here, we still have our amulets of somethingness. We still have matter and energy. True, Albert Einstein famously demonstrated that matter and energy are two ends of the same lucky horseshoe, and that matter is, in the words of the science writer Timothy Ferris, "frozen energy." From tiny quantities of mass we can extract enormous plumes of energy, as the nuclear bombs that destroyed Hiroshima and Nagasaki proved all too darkly. Our sun, too, shines by transforming its core tissue into the pure energy of light and heat; but because it can squeeze so much radiance from so little solar mass, it has shone for 5 billion years and will burn for at least 5 billion more.

Nevertheless, in our workaday world, matter and energy, like the four fundamental forces of nature, behave according to distinct operating manuals, and are proud of their specialized talents. Matter is indispensable to the making of all things — planets, the Crab Nebula, 350,000 species of beetles, four members of the Beatles. Matter comprises the hundred-odd elements in any number of mixes, matches, solids, liquids, or gases. But mass can do nothing of interest without energy. The formal definition of energy is "the capacity to do work," which sounds drearily nagging. Have you finished your algebra yet? OK, then, time to practice piano! Better to think of energy as the opposite of parental or pedantic. Think of energy as romantic. Think of it as a lover, or the idea of a lover, as the spark that makes matter matter. You want to turn on a light. You want electrons to surge through your circuits. They will not move of their own accord. You must excite them. You must supply a source of energy, which will stimulate the electrons in the circuits and send them streaming and screaming to do your bidding.

In thinking about energy, forget for a moment the upsetting image of large oil rigs being constructed in the world's few remaining wildernesses, where the heavy machinery may scar the landscape and disrupt the ecosystem long after having extracted the bare semester's worth of fossil fuels that lie underneath. Consider happier ways to get energy. You can eat a bowlful of cherries, offering your body a source of complex carbohydrates that it can break down into smaller pieces, thus releasing the so-called chemical energy that had held the carbohydrate chains together. You can set up a windmill that exploits the moving currents of air to turn a blade that turns a crank that powers a pump that generates an electric current. Or you can hoist the rigs of your sailboat and let the mechanical might of the wind carry you from one leisure-time activity to the next. You and your lover can snuggle in front of a fireplace and warm yourselves with the "heat energy" generated by the combustion of a selection of seasoned hardwood logs, perhaps kindled by crumpled pieces of newspaper or the old love letters of unfaithful partners past. If you see a cockroach too large and revolting to fit comfortably beneath the sole of your tennis shoe, you can kill it with the "gravitational energy" supplied by the release of a brick from your hand onto the floor.

All of these disparate forms of energy that we describe as chemical, mechanical, heat, gravitational, or hysterical are variations on two mega-categories of energy: stored energy, which is more grandly known as potential energy; and moving, or kinetic, energy. A ripe cherry holds potential energy in its carbohydrate bonds. As those bonds are systematically pried apart by metabolic enzymes in your cells, some of the fruit's potential energy is converted into kinetic energy that you can then use to go shopping for more cherries. A frozen lake in the mountains is a reservoir of potential energy that, when the ice thaws in spring and starts burbling downward, becomes kinetic energy of considerable scenic value. A lit match translates the potential energy of timber into the kinetic energy of hot, dancing flames. Lift a brick, and you essentially inject it with potential energy. Drop the brick, and potential energy quickly expresses itself as a kinetic whop on the greasy auburn exoskeleton of an unfortunate *Periplaneta americana.*

The energy that we call electric energy also has its potential and kinetic guises. Electrons and protons are, as a fundamental feature of our atomic world, drawn irresistibly toward each other. Separate them, and the electromagnetic force will hound them to find some way to rectify the imbalance. Move toward a proton! Fill that hole! What do you think you are, a neutron? The electromagnetic force also urges particles of the same charge to keep a certain distance from others of their kind. Push two like charges unnaturally close together, and they will feel hemmed in, keyed up, anxious to spring away. The electric power on which we are so dependent takes advantage of these particulate impulses in numerous ways. We have batteries in which one set of chemical reactions generates a buildup of excess electrons on one end, while a different set of chemical reactions yields a preponderance of positively charged atoms, or ions, on the other end. Give the opposing charges a chance to mingle, and the resulting burst of energy just may light up the room.

For the electric current to flow, however, it needs a path, a circuit, a conductor, just as the excess electrons you picked up from the carpet required the bridge of your finger on doorknob or on nose of pet in order to reach more positive pastures. A length of metal wire linking the battery's negative and positive poles provides that path. The excess electrons at one end feel the tug of positive ions over yonder, and a concomitant repulsion from the other negative yokels around them. They begin jostling atoms in the wire, which shed some of their outer electrons, which in turn thump the atoms a little farther along, and like a row of clattering dominoes the charge is propelled forward. The potential energy of the battery's chemicals is reinterpreted as the kinetic energy of jostling atoms and electrons, which can be tapped to run a motor or heat the filament of an incandescent bulb until it radiates light, wondrous light, compliant Tyger burning bright.

The electric current that streams from your wall sockets courtesy of your local utility also relies on a pushing and pulling of electric charges along compliant channels. The initial sequestering of positive and negative charges is not easy. It takes work to keep protons from electrons, and work requires energy. A fruit tree needs the sun's radiant bounty to blossom, and a power plant needs one source of energy to spawn the handy electric kind its customers demand. Most power plants in the United States burn coal, gradually converting the substantial sums of potential energy banked in these fossilized briquettes of ancient forests into a river of charged particles tumbling forward in a furious crusade to meet their match. And one of the ways that the transformation from charcoal to sparkle unfolds is through the second half of the force in charge of charge: magnetism.

As Michael Faraday and James Clerk Maxwell determined more than a century ago, the electric force and the magnetic force are intimately, mathematically related. Both physicists were brilliant pioneers in the quest to unify the fundamental forces of nature, an industry that keeps thousands of theoreticians gainfully employed to this day. For his effort, Faraday was awarded not one but two standard units of measurement — the farad and the faraday. As for Maxwell, we honor him each time we utter the compressed term he coined: electromagnetism.

But what is magnetism, and why do you have too much of it on your refrigerator door? How are electricity and magnetism related? They are both, it turns out, very good at fieldwork. They generate fields of their own — magnetic fields and electric fields — and the field of one force can affect the behavior of the other force. To talk about a "field" is another way of saying, action at a distance, or, the pluck doesn't stop here. The earth has a gravitational field, a tugging of other bodies toward itself, which progressively weakens the farther from Earth you manage to fly. Similarly, a charged particle like an electron or a proton has an electric field around it, a personal sphere of influence that projects into space and that either repels or attracts other charged particles. As with a gravitational field, an electric field gets feebler the farther from its source you roam. And, as anybody who has ever played with a couple of bar magnets or with Thomas the Tank Engine trains knows, magnets have distinct fields, too, regions of force that radiate outward from each end of the magnet and either repel or suck closer the ends of the other bar. From whence this rigid animal's magnetism? Whether it is a bar magnet, a classic horseshoe magnet painted silver and red, a lodestone at the natural history museum, or the bendable black backing that keeps your veterinarian's business card on perpetual display, the items we call magnets have the unusual property of spin synchronization. As electrons float around an atom, they also spin on their axes, although "spin" in the curiouser context of Quantum Corner is not exactly like a spinning disco ball or planet; for one thing, it takes an electron two complete rotations to get back to where it started from. Nevertheless, electrons cloud around the nucleus and gyrate on their toes, each producing a tiny magnetic field as it spins. Some may be spinning in one direction, some in another, the end result being that in most atoms, the magnetic effects of these motions cancel each other out. In some metals, though, like iron, cobalt, and nickel, the electron spins can become synchronized, either temporarily or permanently, amplifying those little magnetic fields into one big one. Now you have a magnet, an object that, among other properties, generates a magnetic field, attracts iron and steel, and is keenly responsive to electricity.

Send an electric current through a wire and, depending on which direction its electrons are flowing in, the current can demagnetize a magnet, remagnetize a demagnetized magnet, or turn a nonmagnetic metal temporarily magnetic. The flowing electrons of the electric current affect the distribution of the atoms in the magnet or magnet aspirant, aligning atoms of similar spin and thus magnetizing the material in some cases, or jumbling up clockwise and counterclockwise spinners and demagnetizing the substance in others.

The wheelings and dealings are mutual, and a magnet can set an electric current coursing along a wire with cardiovascular verve. If a copper wire is spun quickly around a magnet, the magnetic field will jostle the electrons in the copper and start them dancing from shell to shell, atom to atom. Add a positive incentive to one end of the wire, and the electron surge will flow fiercely toward it. Power plants often create electric currents by spinning large copper coils inside giant magnets at very high speeds, the rotation driven by a coal-powered turbine engine. The domino wave of hyperexcited electrons in the coil is then transmitted along a lengthy grid of power lines, some of them tucked underground, others fastened onto high-tension towers that loom phantasmically over the highway, like a procession of giant Michelin Men with arms of aluminum lace.

When you flip on your home computer, you divert some of the electric current rowing merrily along the wires in that utility pole outside your place of residence, directing it through the distribution line that feeds your household wiring and allowing it to stimulate the electrons in the computer cord. The kinetic energy pumped into the cord can then be assigned to a task, like activating a tiny motor in the computer's hard drive. Or it can re-separate the positive and negative charges in your computer's battery pack: the business we call "charging" the battery, again to the annoyance of electro-purists, who observe that no new charges are added, but rather the existing ones yanked apart and sufficiently segregated so their eventual reunion will have some oomph. With a comfortable supply of potential energy in your battery and at your disposal, you won't weep should a great static stutter in the sky knock a tree onto your utility pole. The lights may wink off but, lo, the computer still beams, and you can work, work, work in the dark.

The transmission of electric current requires a circuit, a pathway of atoms amenable to bartering charged particles, growing excited in the process, and propelling that wave of kinetic energy forward. Yet electromagnetic energy, that is, electromagnetic radiation, needs nothing to propagate it from here to there. Waves of electromagnetic energy can travel just fine through a vacuum, which is lucky for us, or we'd die of cold and hunger and an unslakable longing for the sun. Because the term "electromagnetism" encompasses so many different concepts — the attraction between electrons and protons in an atom and between socks and sheets in a dryer, the flow of charged particles through a wire and the bilious glow of a fluorescent light bulb — it can be easy to overlook or misconstrue the specific beauty of electromagnetic radiation and the incomparable lightness of light.

Nearly all the energy on which we earthlings rely begins with the waves of electromagnetic radiation that billow forth so extravagantly and implausibly from our sun. We may burn coal to make steam to turn turbines to spin coiled copper to make an electric current to heat and light our house on a winter's night, and we may chemically and pyrotechnically "refine" crude oil extracted from thick layers of mudstone, limestone, and calcium sulfate beneath the Saudi Arabian desert into the petroleum that runs our vehicles, but all those "fossil fuels," those stashes of archaic vegetable matter compacted into dense energy candy through 300 million years underground, were fueled first by sunlight. Plants have the molecular tools to capture solar radiation and put it to use. Plants then become food for others — fast food for the cherry pickers of today, or slow food as the fossil fuels of the future. No matter: the real hero here, the author of every story and the wearer of every toque, is the sun. "When you eat a green, leafy vegetable, you are eating photons of solar energy," said Daniel Nocera of MIT. "You are biting the light of the sun."

We tend to think of sunlight as the light that we see, that the cells of our retina are able to capture and transmit as nerve impulses to be interpreted by the brain. We think of sunlight, in other words, as what we call "visible" light, the small slice of the electromagnetic spectrum that our human eyes can see. Yet most of the sun's light is metaphorically dark, is outside the tiny percent of the electromagnetic spectrum that our relatively impoverished vision can detect. If the sun were a Baskin-Robbins shop with 100 billion flavors on the menu, we would be capable of tasting only 5 of them. We're aware of some of the sun's invisible powers — the thermal radiation that feels warm on the skin, the ultraviolet radiation that makes wrinkles begin. But there are many other species of the electromagnetic spectrum, many ways you can wave and stay light on your feet. Here's a little riddle that Bob Mathieu of the University of Wisconsin in Madison would like everybody to be able to answer. "What do all these things have in common: radio waves, microwaves, infrared, optical, ultraviolet, X-rays, and gamma rays?" he asked rhetorically. "They are all light."

Fine. They are all light. They are all electromagnetic radiation. They are all — what? Electromagnetic radiation is really a couple of big moving fields, one electric, one magnetic, traveling together at right angles to each other. Hard to envision, I grant you, but think of this: An electron is surrounded by an electric field, the charismatic attitude, or force of personality, that other charged particles respond to. If you move a stream of electrons back and forth quickly along a metal conductor, the herking and jerking of their electric fields will generate a magnetic field, which wraps around that conductor the way your fingers wrap around a bicycle handle. The newborn magnetic field in turn provokes the formation of yet another electric field, which then takes up the dare and creates a new magnetic field. Iteration and reiteration, Pete and Repeat, the electric and magnetic fields continue spooling out novel counterpart fields. And as each neonatal field arises, it can amplify, diminish, or otherwise modify the existing fields, depending on whether the fields' peaks and troughs synchronize or interfere with each other. This oscillating blossoming of fields begins rippling outward as electromagnetic radiation — as light. The electrons may be stuck on their trolley line, but the electromagnetic field they fomented can break free and soar through the air or through no air at all, traveling at 300,000 kilometers per second, the universal speed limit at which light has license to fly.

All types of electromagnetic radiation can travel at the speed of light, but they do so with their distinctive style. Depending on how their mutually interacting and propagating fields manipulate one another, light waves may journey in long, gentle swells or compressed, nervous spikes, or any dimension between. You may get a pure signal of like-minded waves or a sampler of shorties, mediums, and XXLs. It is similar to what happens when you drag your hand back and forth through bath water. If you whip up the water haphazardly, peaks and troughs and froths of varying sizes will emerge from the stir. If you get into a cadence of swishing, however, you can instigate a smooth, sinuous, motivated wave, which would surely buoy a rubber duckie outward through all space, time, and divinity if not for the walls of the tub.

Our versatile sun bakes up a banquet of electromagnetic fields and radiates light across the spectrum. But because it is a medium-sized, middle-aged star that, in the stellar scheme of things, is only under moderately high pressure at its core — the source of its electromagnetic glow — a handy proportion of its light is of spritely but not histrionic energy, leaping through space in graceful, compact wavelengths. Those wavelengths happen to lie in or near the visible zone of the electromagnetic spectrum, although "happen to" has nothing to do with it. Our eyes evolved to respond as best they could to ambient light, and the sun is very good at propagating light waves that are between 15 and 32 millionths of an inch long. This is the slice of light that we immodest *Homo taxonomists* have designated as visible light, or optical light, or daylight. Yet the terms are terribly blinkered. Other animals can see light lying well outside the so-called visible range — in the ultraviolet, in the infrared, in radar. Bees, for example, see perfectly well in the ultraviolet range, and many flowers beckon their pollinators with ultraviolet stripes, while the pits of a pit viper detect infrared light, the signature thermal radiation that emanates from meal and menace alike.

Different wavelengths of light are adept at different feats. Radio waves, being very long, can travel without being absorbed or scattered by air molecules, and the longest ones bend readily around the curve of the Earth. They are therefore excellent at transmitting broadcast signals from far-flung radio and television stations to the appropriate receiving device in your home or automobile or, as some people swear, the fillings in one's teeth.

Next down on the electromagnetic spectrum is the ill-named light brigade, microwave radiation. Microwaves are not micro at all, but reasonably wide-bodied, extending from about a centimeter up to a meter in length. Like radio waves, they're long enough to convey signals through the air unfazed. Unlike radio waves, they can be focused into a highly directional beam and hence transmit the signals from one horned antenna to another with a relative degree of security and privacy. Radar is a form of microwave radiation, a directional pulsing of microwaves that reflect off solid objects and back to a receiver, revealing the location of pinged objects with extraordinary precision. A top-of-the-line radar can pinpoint the whereabouts of a housefly two kilometers away, although clearly this is a radar with far too much time on its hands.

Over on the other end of the spectrum, we have X-rays, which are extremely short, about a ten millionth of a millimeter across, or roughly the width of an atom. X-rays are energetic enough to pass directly through most parts of the body but are absorbed by high-density tissues like bone. Yet despite their long-standing utility to medicine, dentistry, biology, and astronomy, X-rays have yet to shed their campy secret-agent of an alias, bestowed on them in 1892 by their discoverer, Wilhelm Roentgen, because he had no idea what they were. Even after the electromagnetic nature of X-rays was elucidated, the quizzical consonant stuck, and by the looks of it will forever mark their spot.

Moving past X-rays, we come to gamma rays, which are pretty much as cinched a wavelength as we can measure. Gamma rays are shorter than a proton's bow tie, but they shoulder massive backpacks of energy. The sun's gamma rays do not make it through our stacked atmosphere without getting lost in the bowels of air molecules en route. Nevertheless, the rays are potentially hazardous to human health and its services. People who fly frequently on long intercontinental flights that cruise through the diaphanous stratosphere six or eight miles above Earth may be exposed cumulatively to undesirable quantities of solar gamma radiation. And should a star located anywhere within about 25,000 light-years from the Earth explode into a supernova, the burst of gamma rays thus unleashed could well knock out entire telecommunication systems. Cell phones, blogs, e-mail, e-dating, e-gads — life as we know it e-rased in a flash.

On the flossy face of it, nature doesn't act like much of a miser. At the end of each spring, forest floors are littered with hundreds of times more fallen blossoms than could ever have borne fruit, and armies of sprouting acorns that will die long before they crown, and the bones of fledgling songbirds that proved surplus and were expelled from the nest. As the brain of a human fetus grows, one hundred neurons must die for every brain cell that settles in and synaptically connects to its neighbors; and the prenate's fingers and toes likewise are whittled down from primordial flippers that fan out from the ends of its limbs.

In its daily sixteen hours of grazing, an elephant eats the equivalent of its trunk's weight in food: three hundred pounds of grass, leaves, roots, bark, branches, bamboo, berries, corn, dates, coconuts, plums, sugar cane, and, as I discovered in my girlhood, Ring-Dings. The elephant's intestines extract only a small portion of the nutrients in that extraordinary intake, though, and the rest is discarded in a similarly astonishing output of about two hundred pounds of dung per day.

Yet beneath nature's extravagant breast lies a thin-lipped bursar, tallying every bean and brain cell, pricing every sheaf of grass. Nature is a tenacious recycler, every dung heap and fallen redwood tree a bustling community of saprophytes wresting life from the dead and discarded, as though intuitively aware that there is nothing new under the sun. Throughout the physical world, from the cosmic to the subatomic, the same refrain resounds. Conservation: it's not just a good idea, it's the law. Isaac Newton discovered some of the laws of conservation. By the law of the conservation of momentum, for example, if a 5,000-pound, all-terrain SUV traveling at 30 miles per hour were to slam headlong into an angry 12,000-pound elephant ring-a-dinging toward it at 25 miles per hour, the relatively more momentous product of the elephant's mass multiplied by its pace would be only partly offset by the opposing but smaller momentum of the moving vehicle, and some of that elephant ire would be transferred to the vehicle, tossing it backward and into the nearest baobab tree.

The law of the conservation of charge means that for every positive charge generated over here, there has to be a net negative charge somewhere else: if, through combing your hair, you turned a few strands into mutually repulsing objects of positivity, you must have infused your comb with extra electrons. You can't snuff out or neutralize a particle's innate charge, and, as far as scientists can tell, the universe is electrically balanced: for every electron there is a proton (turn, turn, turn). You cannot drum up a negatively charged atom or group of atoms — a negative ion — without simultaneously yielding a positive ion.

Perhaps the most profound of all the preservation statutes — and one of two laws of conservation that scientists repeatedly told me they wished the public understood — is the law of the conservation of energy, also known as the first law of thermodynamics. I've long loved the word "thermodynamics," for both its sound and sense of heat in motion. The science of thermodynamics is the study of the relationship between kinetic and potential energy, and heat. The major premise undergirding the discipline is this: in a closed system, the total amount of energy, including heat, is always conserved. Energy cannot be created, replicated, or conscripted from other dimensions. Energy cannot be destroyed, redacted, or forced into early retirement. Energy can only change hands or be converted from one form to another. The qualifying phrase here is "in a closed system." Many systems we encounter in our daily lives are not closed. If you're boiling water on the stove, you can continue adding more energy to the system — that is, the saucepan — simply by keeping the burner on. The kinetic energy released by the protracted combustion of natural gas will be transferred steadily to the water molecules, causing them to bobble about faster and faster until they undergo a phase shift and turn to gas. Even after all the water has boiled away, the energy from the burning natural gas can continue working the system, oxidizing the metal alloys of the saucepan and rupturing the bonds among them, and melting the tough resin polymers of the handle, until finally you, the negligent cook, will need to open another system — the windows and doors of your home, to clear the kitchen of the stench of your favorite saucepan gone to pot.

Other familiar systems, though, are effectively closed — for example, a child on a playground slide. The child climbs to the top of the slide, gathering potential energy in the ascent. The child then sits at the top of the slide, takes a breath, makes sure the parent or guardian is watching with the appropriate mix of excitement and admiration, and lets go, cashing in the stored gravitational energy for the thrill of kinetic energy, along with the inevitable sideline seat warmer of heat. If you added up the kinetic energy of the descent and the energy transferred by heat to the slide, the child's bottom, and the air molecules she rushed past, the sum would equal the gravitational energy with which the transaction began.

The sun and Earth comprise another energy system best thought of as isolated, at least until we invent some version of a *Star Trek* warp drive and can seek out new life and new Saudi Arabias. For the time being, energy derived from solar radiation or its chemical, composted, or meteorological offspring — coal, wood, wind currents — or from the manipulation of matter on Earth in nuclear power plants or the still fanciful fusion rings, will have to suffice.

The system to close all systems, though, is the universe itself. The first law of thermodynamics applies to the entirety of the cosmos. What we have is what we've got and will always have. The energy released at the moment of the Big Bang, 13.7 billion years ago, is our first E, our final E, our only hope and dowry. No deposits, no returns, no spontaneous generation — exchanges for in-store merchandise only. This is not a bad law, really, and in a way it gives salve to the soul. For one thing, there is enough energy to fuel the 10 billion trillion stars of the 100 billion galaxies of the electromagnetically visible universe, as well as the huge quantities of dark matter and dark energy that are lightless and invisible but that we know are out there. Our universe is like a French pastry: full of air yet unspeakably rich, and really, don't you think one will do?

For the nonreligious among us, the law of the conservation of energy offers the equivalent of a spiritual teddy bear, something to clutch at during those late-night moments of quiet terror, when you think of death and oblivion, the final blinding of I. The law of conservation of energy is, in effect, a promise of eternal existence. The universe is, practically speaking, a closed system. Its total energy will be conserved. More will not be created, none will be destroyed. Your private sum of E, the energy in your atoms and the bonds between them, will not be annihilated, cannot be nulled or voided. The mass and energy of which you're built will change form and location, but they will be here, in this loop of life and light, the permanent party that began with a Bang. "Nothing is destroyed, nothing is ever lost, but the entire machinery, complicated as it is, works smoothly and harmoniously . . . the most perfect regularity preserved," waxed the British physicist Joule of thermodynamic's first law. I tell this to my daughter whenever she's scared of the dark, and though she'd prefer a more personalized form of perpetuity, she's found some warmth in this thermodynamic verity. On leaving the house one frigid morning for school, she glanced wistfully at Manny, a purring, well-fed spit curl of fur tucked in the arm of the couch. "After I die," my daughter said, "I hope some of my atoms can find their way into a cat."

Just as the "First World War" carries the suggestion of others and the addition of a qualifying Roman numeral I after "Queen Elizabeth" or "King Felipe" means that others with that name have since been at least nominal heads of state, so the first law of thermodynamics sounds like we're just getting started. In fact, there are four basic laws of thermodynamics — and one of them, conceived after the first, was given the fun-loving name of "the zeroth law of thermodynamics" — but the most important by far are the first and the second. It's like the amendments to the U.S. Constitution. The first guarantees freedom of speech, press, and religion. The second protects your right to bear a howitzer. What more do you need?

As it happens, scientists view the second law of thermodynamics as a firearm of sorts, spraying a scattershot of slugs through the house, knocking pictures from the walls, blowing out the flat-screen television, and making chintzmeat of the furniture. If the first law of thermodynamics is the "good news," as Robert Hazen and James Trefil have written, "a natural law analogous to the immortality of the soul," then the second law is the "bad news," a natural law that helps clarify why the body grows old. The second law might also be called the "Humpty Dumpty directive." Once the big, smirking, pedantic, cravated egg had his great fall, all the king's horses, all the king's men, all the plastic surgeons, duct tape, and members of the National Transportation Safety Board couldn't put Humpty Dumpty together again. The second law is the reason why either you or a hired professional must expend considerable effort to clean your house, but if you leave the place alone for two weeks while on vacation, it will get dirty for free. It explains why some drinks taste good cold, some taste good hot, and most taste lousy at room temperature — red wines, of course, excepted. The second law guarantees a certain degree of chaos and mishap in your life no matter how compulsively you plan your schedule and triple-check every report. To err is not just human: it's divined.

These are some of the philosophical implications and sneaky fine-print clauses of the second law. What are the physical principles behind it? The first, deceptively modest premise of the law is that heat will not flow spontaneously from a cold body to a hot one. If you are carrying an ice cream cone around on a hot day, the ice cream will start to melt. If you continue carrying it around on a hot day, it will melt more, dribbling down the cone, snaking over your fingers, and splattering to the ground. The ice cream will not change its mind and start to freeze up again. On the flip season, if you take your hot coffee outdoors in winter and you don't have a well-insulated mug, the coffee will quickly grow cold. It won't figure out a way of extracting and concentrating whatever small amounts of heat might be had from the moving air around it. In our universe, the spontaneous flow of heat is unidirectional, moving from a warm body to a cooler one. Time and time and time again, the warmth of the summer air will head into your ice cream, and the warmth of your coffee diffuse into the winter snap, and if you suddenly start finding otherwise, maybe it's time to consider rehab.

On a molecular level, heat's arrow makes sense. The molecules of a warm object are moving faster than those of a cool object. As the energetic particles bump into the more stately molecules, they transfer some of their energy to the slow motes, and become less energetic in the barter. The hot summer air molecules bump into the crystals of your ice cream, the crystals start jiggling and break apart; and though the air molecules right around the cone cool down slightly as they convey their verve to the ice cream, it's hard to tell when there's so much hot air to go around. In your hot coffee, the fast-moving molecules at the surface share their energy with the cold air just above them. Those heated air molecules jiggle the air above, while the heat rising from farther down in the cup jiggles the slowed coffee molecules at the surface. In either case, for the transfer of energy from hot to cold not to occur, the cool, slow molecules would somehow have to resist the comparatively greater impact of the speedy, heated molecules, and there is no way for them, of their own accord, to rebuff the body blows.

The result of heat's natural tendency to flow from hot spot to cold is a gradual leveling and homogenization, a diffusion of energy into a limper and less organized pattern. The ice cubes in a glass of lemonade left out on a counter lose their structure. As the heat in a cup of coffee drifts into its surroundings, the molecular reactions that account for the beverage's rich, aromatic zest likewise slow down, and the coffee starts to taste bland. To maintain structure, to maintain a temperature gradient that resists the spontaneous wafting of heat toward coolness, you need an inoculum of energy. You can keep your ice nicely cubed in a freezer, but the intricate cooling mechanism of a refrigerator or freezer is driven by electricity, as are the coolant coils in an air conditioner. You can warm your house in the winter, and counter the gradual loss of that warmth into the frigid air outside, but again you need energy: a wood-burning stove, a furnace fueled by oil or natural gas. No matter how well insulated your home, still there will be a gradual loss of heat to the street, and the consequent call for fresh fodder.

This leads to the second premise of the second law, which can be simply stated as: nothing's perfect. More formally, you can never build an engine that would be 100 percent efficient, able to turn every gram of fuel you feed it into useful, honest, Protestant ethic–approved work. You cannot build a perpetual-motion machine that will keep clicking and tocking without periodic help from outside, though Leo only knows that thousands of humans from da Vinci fore and aft have tried. They fought the second law, the law won. No matter how generously lubricated an engine may be, no matter how beautifully honed and fitted its gears, some amount of the energy that drives it will be lost as heat, will be puffed into the sky rather than turned to the task at hand. Some kinetic energy will end up exciting the air molecules around the engine, or the atoms of the base surrounding the moving parts, or the bolts holding the parts together. Something, somewhere, will take the heat and squander it. Most machines, including all the small, organic ones inside the cells of our body, are far less efficient than 100 percent, or even 50 percent. Many plant species, for example, manage to translate only 5 percent of the energy coming at them from the sun into stored energy to grow on.

To understand the inevitability of inefficiency, think for a moment about a simple part of a car engine, the up-down, in-out motions of the pistons in the cylinders that turn the crankshafts. Each time a piston is thrust down through its cylinder, the air within is compressed and heated. As a result, not only is some of the energy from combustion waylaid into an unnecessary stimulation of the cylinder's air molecules, but now that hot, surly air must be removed from the cylinder before the piston cycle can begin again; otherwise, the engine will overheat and blow a fuse, pop a gasket, disengage from the vehicle's central processing unit, or otherwise cease to function. That demand means opening an exhaust valve to dump the hot air into the atmosphere, where it will abandon all pretense of productive, taxpaying citizenship and instead start mingling with other hotheaded gases that have nothing better to do than disrupt the global climate.

In sum, first you have to pay for something you didn't need or want in the first place, then you have to pay to get rid of it. Sounds like many things in life, doesn't it? Desserts and the gym, the injury you acquired at the gym when you dropped a dumbbell on your finger and the doctor's bill for sewing the finger together again, your daughter's pet African bullfrog and removal of your daughter's decomposing African bullfrog from under her bedroom floorboards. Nothing's perfect, nobody's perfect, and the smart ones don't waste too much time trying to be.

Which brings us to the third and potentially most depressing premise of the second law: every isolated system grows more disordered with time. Or, as a sign on my editor's door put it, ENTROPY ALWAYS GETS YOU IN THE END.

The word "entropy" has gained a certain popular cachet, and is often used as a synonym for chaos, but the two terms have distinct meanings. In physics and mathematics, chaos refers to systems like the weather, or a nation's economy, that seem random and unpredictable but that often have regular, repeated patterns underlying them — high pressure clouds, the PBS broadcasts of Suze Orman. Entropy, by comparison, is a measure of how much energy in a system is "not available to do work." The energy is there, but it might as well not be, like a taxi passing you on a rainy night with its NOT IN SERVICE lights ablaze, or a chair in a museum with a rope draped from arm to arm, or a teenager. Rudolf Clausius, a German physicist and thermodynamics pioneer, coined the term "entropy" from the Greek word for "transformation," and he coined it with care, to sound as much like "energy" as possible. Wherever you have energy, Clausius said, entropy is sure to follow, with crowbar in hand.

The first law of thermodynamics insists that energy can be neither created nor destroyed. The second law replies, Fine, then I'll have to settle for breaking its knees.

In a closed system, entropy creeps higher, and order slowly subsides. It is a cold, hard, tepid, flaccid, probabilistic truth. If you bring a pot of water to a full boil, put an egg in it, cover the pot with a tight-fitting lid, and turn off the flame, you'll have a reasonably isolated system. The water will be hot enough to cook the egg to a soft-boiled or medium-soft consistency. But at some point before the egg reaches the child-friendly status of maximum firmness, the system will lose its culinary power. Much of the kinetic energy that the water molecules had won through boiling will be shrugged off as heat into the air under the lid. In their less vigorous state, the water molecules abutting the eggshell cannot continue revising, linking, and cross-linking the egg proteins and cholesterol chains within. The total energy of the system may be the same as it was when the lid first descended, but it has become diffuse, and defused — it's no longer cooking with gas.

The second law, alas, has overwhelming odds in its favor. When physicists speak of an ordered system, they mean one in which the components are organized in a regular, predictable pattern, as the atoms of sodium and chloride are neatly stacked in a crystal of salt, or as books are arranged in a meticulously managed library — thematically and alphabetically. But think of that library, and how easy it is to perturb its order. You don't have to reduce the entire collection to a jumble on the floor; a single, misinserted volume is enough to ruin a scholar's whole morning. In fact, there is only one way for the books to be arranged on the shelves in a flawless Dewey-decimal sequence, but thousands upon hundreds of thousands of ways that they can be set astray. Herein lies the engine of entropy. Order, by definition, has restrictions and limitations, while disorder knows no bounds. The odds of the boiled water in our pot retaining its heat by dint of the agitated water molecules on the surface repeatedly bumping into only other water molecules below and beside them, rather than some of them slapping against the air molecules above, are infinitesimally small. In theory, it could happen, just as you theoretically could close your eyes, begin tossing a couple of hundred bricks into the corner, and find, on opening your eyes, that you've thrown them into the perfectly aligned, beautifully crafted Flemish-style hearth wall of your dreams. In probabalistic reality, you will be treated to the sight of a haphazard pile of half-cracked bricks and a haze of pulverized clay, and to the sound of the police pounding on your front door with heavy objects of their own.

No, if you want that ton of bricks to shape up into a presentable mantelpiece, you'll have to get out your trowel, bucket, and mortar, and you'll have to invest some of your stored chemical energy arranging and rearranging the bricks and daubing over the rough spots, sort of the way evolution did with us. You can also count on the need for periodic touchups and repointings, as the impact of heat, gravity, dankness, cold, grease, pine tar, mold, the rattle of passing garbage trucks, the time you called in the unlicensed chimney sweep because you didn't realize you had to open something called a "flue," all nudge the wall's brick and cement molecules into predictably shabbier configurations. Eventually, you or one of your descendants may decide that the wall is splintered and sagging in so many places that it's easier to get out the sledgehammer and start all over again.

By the second law of thermodynamics, the energy of a system may remain the same in quantity, while steadily declining in quality. The concentrated energy of petroleum is quite useful; the dispersed energy of excited molecules of carbon dioxide and nitrogen oxide belched out of a car's exhaust pipe is not. The darkest readings of the second law suggest that even the universe has a morphine drip in its vein, a slow smothering of all spangle, all spiral, all possibility. In this version of the apocalypse, universal entropy is rising, productive energy falling, and the entire package fading into cool irrelevance. Today, the explosive death of one star can infuse a nearby gas cloud with so much energy and matter that it collapses into a brand-new baby star, and the stellar life cycle pedals on. In the bigger and more estranged cosmos of the distant future, there may be no suns left with the will to explode, or nurseries to seed with their heirlooms of light.

But before we get carried away by a formaldehyde gloom, let's remember that, whatever its eventual fate, the universe still has an awful lot of time left to play, and that it is a comic genius and an aesthete that defies its innate sloth, its entropic drift, with sustained symphonies of disciplined beauty. The universe loves patterns, and it can't seem to stop finding new styles of light and character, and functional forms and dysfunctional forms just for the fun of it. From formlessness came the cloud of glory we named atom, from ashes and dust came stars so formally formed that we can tell by their light how long they will shine and when and how they will die. Atoms were not content to stay in their element, as lonely elements, but instead linked arms with other elements, becoming the molecules of which our world is forged, and the chemistry was right to scoff in the face of the law, and declare, Let's go get a life.

5

Chemistry

Fire, Ice, Spies, and Life

THE NEXT TIME you think you are being teased, picked on, and put upon, don't bother getting defensive. Don't give your tormenters the pleasure of your petulance. Instead, why not try the chemist's solution, and fight fire with . . . a party trick?

Many chemists admit to the occasional indulgence in a persecution complex. They feel demonized by the public and marginalized by other scientists. Chemistry is the subject that at least 6 out of every 6.0225 Americans insist they "flunked in high school." The boilerplate evil scientist of Hollywood is often some type of chemist, a white coat cackling over his boiling beakers and crackling gadgetry. People rant against all the "chemicals" in the environment, as though the word were synonymous with "poisons." The environment, chemists counter-rant, is nothing *but* chemicals, and the same can be said for us. "We're just self-replicating carbon units, that's what we are," said Donald Sadoway, a professor of materials chemistry at MIT. "We're not a heck of a lot different than the carbon-based fiber in a steel-belted radial tire, so maybe we shouldn't take ourselves too seriously."

When not feared as a threat to air, water, fish, and fowl, chemistry is belittled as bureaucratic, a field neither fish nor fowl. Roald Hoffmann, a chemist and poet-playwright at Cornell University, has observed that because chemistry is "poised between the physical and biological universes" and "does not deal with the infinitely small or large," it may be thought fussy and tedious, "the way things in the middle often are." Some of the most heedless squeezers on chemistry's domain are those passengers seated on either side of it. "Chemistry is the core science, the central science, yet its contributions are often overlooked, including by many biologists, physicists, medical researchers, and others who should know better," said Rick Danheiser, a chemistry professor at MIT. Even the much discussed book *The End of Science,* which argued that the major scientific disciplines were reaching the limit of their explanatory power and would soon be irrelevant, didn't bother mentioning chemistry. Danheiser sighed theatrically. I guess we're not even sexy enough for an obituary, he said.

Time for that party trick. Danheiser, a boyish-looking baby boomer with a build that could be described as neither infinitely small nor large, and who balanced the casual bohemianism of his beard with the casual preppiness of a polo shirt and khaki slacks, pushed his chair back, and began rummaging through the drawers of his desk. No luck. He walked over to his bookcase, scanned the shelves, and ran his hands over the edges of those above eye level. Still nothing. Finally he asked me if I happened to have any matches on me. I told him, No, I'm not a smoker. "Well, that's good, as far as your health is concerned," he replied. "But it's too bad for the point I want to make." A round of charades would have to do.

Danheiser took a stick of plastic of the type that chemists use to construct stick models of molecules. "Pretend this is a match," he said of the sticklet, and I nodded. "This," he said, lifting the sticklet over his desk, "is physics." He let the stick drop on his desk. Plink. "And this," he said, scraping the stick's virtual phosphor on a virtual matchbook cupped in his hand, "is chemistry." He held up the stick triumphantly for me to gaze on the fabled flame. It was a good thing that I could summon up an image of combustion well enough to smile and nod appreciatively at Danheiser's act because two other chemists went through similar matchless presentations with me, to make the point about the soul of their discipline. Chemists may feel unappreciated. They may be thought by many adult survivors of a high school education to have the sex appeal of a cold sore. Yet chemists know that through it all they can claim Prometheus as their prophet, and that if they can't find a match in their pocket, there's always the fire inside.

And while we're on the subject of myths and fantasies, a couple of other legendary figures are relevant to chemistry's specific alchemy. One is Goldilocks, whose story offers a plangent alternative to the cliché of chemistry as gray middleman. For Goldilocks, it is the extremes that prove stiff and humorless, the extremes that can never suffice. Too hot will strip the epidermis from your tongue, the electrons from your atoms; too cold, and you can't taste a thing. Too hard, you're not alive; too soft, you've already died. Goldilocks's preferred habitat, her optimal culture medium, is that of the harmonious compromise, the world she calls "just right." It is a world fit for children, human and ursine alike, a world suited for growth, for the calibrated assimilation of atoms into molecules, molecules into compounds, units into chains, chains into pleats, folds, tissues, organs, eyes, snouts, and mouths that shout. A world that is safe for children is a world of immense molecular diversity, where a million molecules and compounds bloom, and where no molecule is left unclaimed for very long. The chemist's world is the world around us, a pampered stratum of relatively mild temperatures, and manageable atmospheric pressure, and liquid water in abundance to bring molecules together and lubricate their discourse.

Roald Hoffmann said, "There is no chemistry to speak of on the surface of the sun. It's all atoms and ions — atoms that have been shocked apart."

On Earth, under our conditions, we have lots of chemistry. We have temperature conditions where molecules can exist in three different states, as solid, liquid, or gas, and where, with the input of energy, from the light of the sun or the heat of a fire, those molecules can change into other molecules, into other complex assemblages of atoms. "What a dull world it would be if there were only 115 of us, and what a dull world it would be if there were nothing more than the 115 elements, 115 different types of atoms, end of story," Hoffmann said. "But that isn't how our world works. From the 115 elements you can build a near infinity of molecules, of any type you need, to get all the structural and functional diversity you can ask for. There are at least 100,000 different molecules in the human body. Some 900 volatile aroma components have been found in wine. Chemistry is molecules. We are molecules. Chemistry is a truly anthropic science."

On the surface of the sun, where temperatures hover around 10,000 degrees Fahrenheit, the atoms are in shock, yet they are not alone. All around them are other atoms, most of them hydrogen, but with a decent number of helium atoms as well as a scattering of carbon, nitrogen, oxygen, and neon, to name a few. What is the difference between the atoms that congregate on the face of our sun, and those that compose the face of our daughter? What must atoms do en masse to qualify as molecules, what password must they utter?

"The name is Bond, James Bond." said Donald Sadoway. "Chemistry is all about making and breaking bonds." In his impeccably tailored Italian suit, crisp white shirt, and elegantly vivid tie done in an expert double Windsor knot, Sadoway had a distinctly Bondish beam himself. Molecules, he said, and the broader category of compounds and mixtures, are more than cohorts of atoms that all happen to be in the same vicinity, like passengers on a train or marbles in a box. To merit designation as molecules or chemical compounds, the constituent atoms must be stuck together with some sort of electromagnetic glue. The atoms must share their outermost electrons with one another, or must feel the persistent tugging of an oppositely charged atom by their side. Chemistry is about molecules, and making bonds and breaking bonds. Chemical bonds are forged of electromagnetic forces, of the innate attraction between electrons and protons, and the fickleness of electrons, the willingness of those negatively charged particles to list one moment toward this proton, one moment toward that. Chemistry exploits electron restiveness to snap the hundred-plus elements of the periodic table into hundreds of thousands of configurations, and to break the bonds apart again and rearrange the pieces again and restock the shelves with new and improved molecular merchandise. Brighter whites! Nightlier darks! Sweeter smells, stronger laminates, longer polymers, snappier comebacks. Whatever your chemical demand, whatever shape, size, or attribute your molecule must possess, chances are you'll find it somewhere in the well-stocked toy box that is our Goldilocks world — if not prefabricated naturally, then conjured in a laboratory. Roald Hoffmann has called chemistry "the imagined science"; the great nineteenth-century French chemist Claude Berthollet declared it an art. "Chemistry is almost alone among the sciences in its ability to make new things," said Stephen Lippard, a professor of chemistry at MIT. "Beyond studying the world as it exists, we can put together combinations of molecules in ways never dreamed of before." A computer screen so flexible you can roll it up like a newspaper and stash it in your pocket; a windshield that cleans itself; artificial arteries that don't get clogged and that the immune system won't assault; antidepressants that conquer despair without the standard civilian casualties of fat and frigidity. Such is the stuff that chemists' dreams are made of.

And as in sleep or art, it is not always clear who is the dreamer, who the dream. "My field is material chemistry, and one thing we don't admit to young students is how clueless we really are," said Frank DiSalvo, a professor of chemistry at Cornell University. "Much of what we come up with we happen on by trial and error, and we can't predict what we'll get ahead of time. We just don't know the rules of the game for more than a handful of the elements we work with." In theory, he said, all the material needed to construct any device imaginable, a warp drive, a transporter, the perfect toupee, is already there, somewhere in the periodic table. It's figuring out where it is, and with whom it should consort, and under what conditions, that keeps the midnight Bunsens burning. "If every person on the planet were a materials chemist," DiSalvo said, "it would still take a millennium or longer to understand the periodic table well enough to make all the things we want to make."

The basic themes of chemistry are molecules and the bonds that bind and define them. As with the lovable British hit man of twenty-odd films, there is more than one way of being a Bond. You have your suave, supple, catlike bond, your stiff-shanked bond, your uncommitted, barely there bond. The type of bond that links together atoms in a molecule, or one molecule to another, explains why the carbon lattices of a diamond are hard enough to be a girl's best friend forever, while the carbon chains in our food are broken down with only moderate metabolic effort, and the carbon molecules on the graphite tip of a pencil can be transferred onto paper using only the most feathery of strokes. Bonds are stirred, bonds are shaken, bonds, like rules, are made for breaking.

The strongest and simplest but by no means most simple-minded bond in nature is the covalent bond, when two atoms team up and share a pair or more of electrons for the sheer goose-down comfort of it. The bond arises between players in a similar state of discretionary desire: their outermost shells don't really need extra electrons, but they have room for them anyway. The individual atoms theoretically are capable of electromagnetic self-sufficiency, the number of orbiting, negatively charged electrons balancing the number of positively charged protons within. But the orbital paths or shells along which electrons travel as they circumnavigate an atom are designed to accommodate a set number of negative particles apiece, the atom's particular proton needs notwithstanding. Orbital shells, in other words, are a lot like closets: happiest when filled.

Atoms with a moderate degree of empty shell space often end up consorting together, and satisfying each other's closet cravings by swapping their outermost electrons back and forth, back and forth. In that way, they get the sensation of orbital satiety without becoming formally, electrically charged, as they would if they picked up one too many electrons full-time, or if they lost the members of their partially filled outer shell altogether. The shared electrons may sometimes be found closer to the outer shell of one atom, at other times nearer to the cloud around the other, but more often fluttering somewhere in between.

"On the one hand, the two atoms want to come together, because their shared electrons want to feel the effects of both positive nuclei," said Roald Hoffmann. "On the other hand, the nuclei don't want to get too close to each other. The compromise distance is the bond length, and it acts as a kind of spring linking the atoms together." Boing, boing, please do as I say, first you come closer, now hop away.

The participants in this covalent Slinky sling may be atoms of the same element. A couple of hydrogen atoms, for example, each with their lone electron in a shell built for two, may pool their particles covalently to form a molecule of H_2, while the oxygen we breathe consists mostly of O_2, vaporous plumes of covalently twinned oxygen atoms that share not just one but two pairs of electrons per partnership.

Alternatively, a covalent vinculum may clasp together two entirely different elements into a so-called compound. Hydrogen and its only-child electron can be hitched to chlorine, which has seven of its seventeen electrons fluttering around an outer shell suited for eight, to form the familiar chemical, hydrogen chloride, a colorless, suffocating, corrosive gas used in making plastics and many other industrial operations. Nitrogen, with five electrons in its outer shell, and oxygen with six, both have outer orbital capacity for eight electrons and can join forces in a variety of permutations. One nitrogen covalently bonded to one oxygen gets you nitric oxide, NO, a clear, potent gas that is quite toxic in large quantities but that the body exploits judiciously for tasks like relaxing muscles, battling bacteria, sending signals in the brain, and engorging the genitals during sexual arousal. In another magic merging of nitrogen and oxygen, a covalently packaged pair of nitrogen atoms can be induced to fraternize covalently with a unit of oxygen, yielding nitrous oxide, N_2O, a sweet-smelling psychoactive gas that makes dentistry almost affable, if never truly laughable. The carbohydrates in our diet are covalently bonded armadas of carbon, hydrogen, and oxygen atoms — carbon and water — the exact proportions and positioning of each element in a given array determining whether the carbohydrate is complex and nutritious or sugary and suspicious.

As a rule, elements are more stable and less chemically reactive when they're in a bonded relationship than when they're out of one, for the same reason that married people are celebrated as society's source of levelheaded bourgeois dependability. When you are married, your coupling capacity is more or less filled, and you are considered "taken." Not for nothing is the emblem of marriage, the wedding ring, a closed circle. Similarly for chemical partners in bondage: their reactive parts are already busy and so are unavailable for other relations.

Molecular marriage does not, however, demand monogamy. Many elements have more than one reactive option, more than one electron consigned to life in a half-full orbit and thus in a position to conjugate covalently with another atom. Many elements, then, are polygamous by nature, and each has its romantic limit, the maximum number of partners with which it can conjoin simultaneously. That figure is known as the element's valence number, from the Latin word *valentia*, for "power" or "capacity." The closer an element comes to filling all its gaps, the more stable, the less chemically predatory, it becomes. The reason why nitric oxide is such a prickly chemical is that, although its nitrogen and oxygen components are covalently linked, both still have room for more electrons and will readily engage in supplementary affairs or frank acts of larceny. Nitric oxide is particularly deft at stealing electrons from iron atoms at the core of hemoglobin molecules, disrupting hemoglobin's ability to convey oxygen throughout the body.

In the case of nitrous oxide, by comparison, all three of nitrogen's available outer electrons are fully engaged in covalent liaisons and are not open for further chemical dalliances, making laughing gas a reasonably benign compound when used in moderation. Nevertheless, the persistence of reactive prospects on the oxygen end of the coalition means that nitrous oxide also can disturb hemoglobin performance, and if you breathe the gas too long you will suffer gradual oxygen depletion and eventually laugh your last.

Nitrogen on its own is capable of extreme stability. In the absence of any pressure to bond cross-culturally with oxygen, hydrogen, or the like, two nitrogen atoms will readily fulfill each other's every need by sharing all three pairs of their available electrons. This triple-bonded nitrogen duet makes for an exceptionally doughty and unreactive molecule that lasts and lasts, which is why liquid nitrogen is the chemical of choice for long-term storage of such prized biomedical goods as blood, sperm, fertilized embryos, evidence from a crime scene. About 78 percent of our atmosphere consists of triple-bonded nitrogen gas, compared to the 21 percent assigned to oxygen; but while our lungs are designed to extract that oxygen from the air and put it to work in every cell of the body, and we are incapable of living without oxygen for more than a few minutes at a time, the nitrogen we inhale is of no use to us physiologically, and we either exhale it immediately, or excrete it later as waste. The nitrogen we do need, for our cells and our DNA, we obtain from food, where the nitrogen arrives in a form that has been conveniently "fixed" for us, that is, combined with oxygen and hydrogen, by compliant microorganisms in the soil; those microbes had "fixed" the nitrogen from the air and fed it to plants, which in turn fed it to us or to the animals we eat. From wherever on the food chain we can pinch it, this molecularly domesticated form of nitrogen is essential to our perpetuation, and all of us can be said to have a nitrogen fixation.

Yet what feeds life can also seed annihilation. Triplex nitrogen is generally unreactive, and its three covalent bonds are very difficult to break, but with the right chemical or incendiary maneuvers they can be broken; and upon rupturing they release large amounts of energy into their surroundings — in short, they go boom. For that reason, most explosive mixtures, including dynamite, gunpowder, and the stuffing in your average nonnuclear bomb, contain some form of thrice-bonded nitrogen.

Chemistry is about molecules, and the word "molecule," like so many scientific terms, has its precise and its casual definitions. Its meticulous meaning is a group of atoms linked together by covalent bonds, by a sharing of pairs of electrons. Yet even chemists sometimes dispense with the formalities and call any sort of chemically bonded substance a molecule, offhandedly referring to molecules of table salt, for instance, or the molecules of magnesium bromide in a bottle of milk of magnesia. In truth, sodium chloride, magnesium bromide, calcium chloride and the like are not molecules but ionic compounds, and though the hero here is still a bond, Sean Connery it is not. The ionic bond that brings us condiments, pebbles, eggshells, Alka-Seltzer, many household cleaning products, and a surprising selection of psychiatric drugs, is stiffer and more strait-laced than a covalent bond, less pliable, more predictable. A brick, a rock, the salt of the earth. An ionic bond is Roger Moore.

In contrast to a covalent bond, which can join together atoms of the same or different elements, an ionic bond can only assimilate the dissimilar. The reason for that is embedded in the term: an ionic bond is a bond between ions, or electrically charged atoms. It is the attraction that a negatively charged atom, laden with one or maybe more electrons than its proton content calls for, feels for a positively charged atom, one that has too few electrons to suit its nuclear desires. Some elements are quite prone to becoming negative ions, others to having an electron stripped away and leaving them positive, but no element is in jeopardy of both ionization fates. When ion-plus seeks ion-minus, you know there's no chance of incest.

The elements at greatest risk of electron loss are those with a single or maybe two electrons in an outer shell intended for throngs. Several inner layers of electron shells separate the outlier from the positive charges in the nucleus. A glancing blow, a brisk breeze, a winking neighbor, and, whoops, the electron's gone.

By contrast, elements likeliest to turn negative are those whose outer shells are practically filled, but there's room for one more. Sure, the element can and often does enter into a covalent time-share, but, Oh, the temptation to go further: just one more electron, one little extra charge, and the entire house would be occupied in earnest, and how wholesome, how aesthetically gratifying that would feel. Just one last little after-dinner mint . . .

Consider, then, the lovely symmetry of salt. On one side we have sodium, a soft metal with the silvery sheen of herring scales. Sodium has eleven electrons, two in the innermost orbit, eight in the next, and, in orbit number three, a solitary sailor with a distinct propensity for jumping ship. Across the aisle, we see chlorine, a corrosive, greenish yellow gas. The outer shell of chlorine, as I mentioned earlier, is one electron shy of satiety, and so chlorine leans toward mean, toward stealing electrons where it can. You can't eat pure sodium, and you shouldn't breathe pure chlorine: they're both toxic. Put the two together, though, and enjoy the show. In a fiery reaction, the sodium atoms essentially wilt and shrug off their extra electrons into the palms of their chlorine counterparts. The sodium atoms in the sample are now electron-deprived and positively ionized, while the chlorines, in fully staffing their orbits, have turned negative (which grants them a name change to "chloride"). Now the two elemental tribes truly want each other. Now the sodium and chloride ions are drawn closer not by the middling desire to round out their shells, but through the much stronger draw of electromagnetic attraction.

At the same time, we have two competing pressures: the attractive tug opposites feel toward each other, and the repulsive sensation between the like-charged ions. As a result, the atoms quickly settle into a regular alternating pattern of chloride and sodium atoms. They stack up neatly in three dimensions like a balanced composition of oranges and grapefruits. These repetitive, geometrically elegant atomic arrays are crystals — salt crystals. What before were two substances that you wouldn't have fed to your old home economics teacher even after she gave you a C for sewing your apron pocket on upside down have condensed into a seasoning so precious that wars have been fought over it, and soldiers given money specifically to buy it — hence the world "salary," from the Latin *salarius,* a stipend for salt. If you were to look at some table salt under a microscope, you'd see just how crisply Pythagorean the grains are, like a sprinkling of art deco glass bricks. Bear in mind that each one of those crystals is an ensemble of a billion billion chlorine and sodium ions, more atoms per granule than there are stars in the Milky Way. Now, would you please pass the salt?

Yet another sort of atomic bond is the metallic bond, the almost socialist sharing of electrons among many atoms in, for example, a piece of copper wire, or the gold of a wedding ring, or the soft sodium sample before its encounter with chlorine. In a metallically bonded substance, the outermost electrons float about in what's often called an "electron sea," being tugged first toward one atom, then toward another, their fluidity accounting for a metal's capacity to conduct an electric current.

The bonds that bind atoms and ions together are all fairly strong glues, with the result, Roald Hoffmann has written, that under normal, nonsolar conditions, "the atoms cohere, move as a group." They cohere covalently as molecules, or ionically as salts, or ironically as metals. Beyond the coherent cliques are larger assemblages, gangling groups of molecules or ionic compounds that adhere together through a couple of bonds of their own. The two big-canvas bonds are weaker than those that marry atoms into molecules, yet they have proved indispensable to life, and ships, and sealing wax, and they give a pencil wings.

One of the critical cross-connectors is the hydrogen bond. The name is unfortunate, not only because it sounds uncomfortably close to "hydrogen bomb," but because it suggests a bond that links hydrogen to other atoms — to oxygen in H_2O, for example, or chloride in hydrogen chloride. The bonds in those cases are, however, covalent bonds, and they are far more serious than a hydrogen bond. In fact, the hydrogen bond is best exemplified by the stridently unserious image of Mickey Mouse: a big round head with two round ears on top. Mickey Mouse here is a molecule of water, with the head representing oxygen, the ears the two hydrogen atoms covalently linked to it. Fortunately, we can dispense with facial details and avoid the risk of copyright infringement.

It turns out that the electron pairs binding each hydrogen ear and the oxygen skull are not quite fairly, squarely, and roundly shared. They tend to spend a bit more time near the oxygen nucleus than near the proton core of either hydrogen atom. As a result, the ears of the Mickey molecule have a slight positive charge: their protons are not always fully counterbalanced by a constant cloud of negative charge. At the same time, because the oxygen atom is hogging a bit too much of the shared electrons' attentions, the bottom half of the mouse face has a five o'clock shadow of modest negative charge. The molecule is polarized; its distribution of charges gives it a directionality, an upside and a downside.

What happens when you put a whole lot of polarized Mickey Mouses together in one place — like, say, Lake Michigan? The chins of one molecule are drawn gently toward the ears of another, lending water an overall shape and integrity that make Mickey quite mighty. Through the puzzle-piece fusing of tops and bottoms, hydrogen bonds account for water's exceptional clinginess, the tendency of droplets to stick together and trail one another loyally no matter where their scout leaders may venture. Hydrogen bonds are only about one-tenth as strong as covalent bonds, but what they lack in strength they make up for with elasticity. Because of hydrogen bonds, plants can drink water; even the crowns of towering redwoods can be quenched. Slender threads of water snake upward from the soil and through the plant's vasculature, to escape as water vapor via pores in the leaves. And as the leading edge of the water column evaporates into the air, hydrogen bonds pull up more fluid from below.

Yet water's hydrogen bonds are slippery and will slide aside to stir things up when something thicker this way comes. Water has been called the universal solvent, for there are precious few substances that will not dissolve in its embrace. Stir a spoonful of salt into water, and water's mighty mice swiftly interpose themselves between the individual crystals, the positive ears tempting the negative chlorines, the negative jaws jockeying for sodium, until the salt grains have disintegrated into a fine mist. Give polarized water molecules about 6 million years, and they'll squeeze blood red beauty from stone, chipping 6,000 feet deep and 277 miles wide into Arizona's northern plateau, through limestone and sandstone and iron-rich shale, to scoop out a canyon the whole world can call Grand.

Hydrogen bonding, though, is not unique to water. It arises in other cases where hydrogen, the lightest of the elements, enters into a covalent compact with a bulkier element, such as nitrogen, and the shared electrons tip their allegiance toward the nucleus of hydrogen's partner. From that asymmetry, a molecule that is electrically neutral in totality assumes a Mickey-like fuzz of charge around the ears and chin.

Another intermolecular melder is the van der Waals force, christened after the Dutch physicist who discovered and mathematically characterized it in the late nineteenth century. Despite the intimidating length of its name, van der Waals is the weakest of the links, as anybody who has ever left a pottery class streaked in enough clay for a solstice fete can attest; its strength is less than a quarter that of a hydrogen bond. Still, a mild manner has its advantages, and van der Waals is essential to the integrity of many solids and liquids, and to the properties of a wide assortment of substances on which we depend. Whereas in other bonds, including the hydrogen bond, electrons tend to know their place and to accord the resulting molecule or compound a fairly fixed arrangement of negative and positive charges, the van der Waals force showcases the electron's improvisational skills.

Electrons, of course, don't like the feel of other electrons, and that antipathy explains why we can touch objects constructed of the near nothingness of atoms and not go right through them. At the same time, electrons are drawn toward protons — the positive particles of their own nucleus or of any nucleus in the neighborhood. The same predispositions apply to electrons in the corporate environment of a liquid or solid, when they are part of molecular or ionic teams. Nuclear protons, good; other electrons, bad. The consequence of this fundamental preference is that, when atoms and molecules come into close proximity with one another, their electrons tend to shift themselves to one side of the home cloud or another, avoiding regions of ambient electron glut, seeking out patches of heightened proton pull. The molecules hence become ever so mildly polarized, or asymmetrically charged, and this gentle layering of negative and positive charges helps bind many substances together. It's a frail fraternity. Electrons are not formally shared between atoms, as they are in molecules and ions, nor are they committed to their unbalanced orbits, as they are in the Disney design of a water molecule.

Still, van der Waals is sometimes the only force holding big chunks of matter together. Pottery clay, for example, consists of sheets of diverse atoms — silicon, aluminum, oxygen, hydrogen, calcium, nitrogen, iron, maybe a sprinkling of cobalt, copper, manganese, and zinc. Within each sheet, the atoms are lashed together by persuasive covalent and ionic bonds. But between the sheets, only the van der Waals force can be found. That's why it's so easy to smear a bit of the putty onto your fingertip; all you're doing with the pickup is interrupting the circumstantial attraction between sheets of mildly polarized clay particles. The integrity of the molecular bonds themselves becomes obvious, though, as you struggle to remove the fine clay slick from your fingertip and discover that it is very difficult to disengage or smoosh apart. Hours later, you may still feel a greasy residue clinging to every whorl — the lingering clay molecules that can be removed definitively only by fracturing their covalent bonds with a serious detergent or chemical solvent.

Your ordinary, standardized test–taking pencil offers another instance of the sound of a van der Waals snapping. The "lead" of the pencil (graphite was long thought to be a soft form of lead, and by the time chemists realized otherwise, the term "lead pencil" was already lodged deep in the schoolhouse lexicon) is not lead at all, but graphite — countless sheets of carbon atoms stacked one on top of another, rather like the tissue-fine layers of caramelized toffee inside a Butterfinger candy bar. Within each graphite sheet, the carbon atoms lock elbows covalently in repetitive, crystalline patterns, but van der Waals alone joins one leaf to those above and below. Press the point of the pencil on your paper to fill in your oval, and you shave a stratum or two of carbon crystals away from the larger deck.

From this versatile cast of bonds, all the stuff of life and site can be staffed. Ionic bonds characterize much of the glittering landscape on which we and our steel-belted radials tread — the mountains, hills, rocks, sand, the shattered seashells by the seashore, the bleaching coral reefs beneath. Ionic solids tend to be rigid, their ions bound together so stiffly they cannot easily be pushed aside. That rigidity makes ionic solids ideal for load-bearing tasks: what better way to begin a bridge than with a few ionically bound concrete pylons, and what better batter for a sidewalk than an ionic compound like cement? Our skeletons, too, are made in part of ionic solids, tightly interdigitated concatenations of calcium, phosphorus, and other atoms. Through bone, we clawed free of the bog; we carry our steppingstones inside us.

Yet ionic solids will get you only so far, and their strength tends to be brittle. You can push down on them and they'll hold up stoutly, but give them a few good rotations, or maybe one swift blow with a hammer, and their ionic bonds will rupture and the crystal palace crumble. For that reason pylons are buried underground, to stabilize them; an uppity tree root can buckle a sidewalk so badly the panel will crack in half; and the twisting of an ankle may end up fracturing the bone. Luckily for us, our bones are marbled through with a soft mortar of proteins, which gives them far more flexibility and torque resistance than they'd have if they were nothing but ionic columns. And lucky, too, that beneath the most brittle outer sheath of our bones lies a network of regenerative tissue that can give birth to new bone cells, sealing cracks and healing breaks and doing for our vertebrate frames what could not be done for the ionic solid of an eggshell: put us together again.

Most of our body tissue consists of covalently rather than ionically bonded compounds — of molecules, not salts. We're abundantly hydrated, of course, and can blame at least 60 percent of our body weight on water molecules, more if we're a pedestrian in Manhattan too embarrassed to try sneaking into a restaurant's FOR CUSTOMERS ONLY bathroom when the maître d' isn't looking. Wring us out — finally! — and the bulk of what's left explains the sci-fi canard about humans as self-replicating carbon units: about two-thirds of our dry weight consists of carbon. Water may be the solvent of the universe, but carbon is the duct tape of life. Every cell, every component of the cell, is based on carbon. If you're seated somewhere on the tree of life, or are the tree of life, *ipso facto* you contain carbon, and that goes for bacteria, amoebas, lichen, dust mites, pinworms, creationists. Even viruses, considered by many to be less than certifiably alive, nonetheless contain carbon, as part of the genetic backpack they tote around from host to host. Small wonder that half of all chemists work in the field of organic chemistry, which has nothing to do with the pure 'n' natural foods industry but instead is the study of compounds that contain carbon.

We are "just" carbon-based units because carbon makes for a just-right class of molecules. Carbon is strong, resourceful, flexible, sociable. With its outer shell of four electrons and four electron slots for rent, carbon is supremely suited to molecular bondage. It happily collaborates with nearly every actor on the periodic table, save helium, neon, and the four other noble elements,* so-called for their snobby refusal to connect chemically to anything. Moreover, carbon is unparalleled among elements in its ability to join with itself almost indefinitely, forming carbon chains and carbon loops and branching carbon prongs and broad carbon planes and bouncing carbon buckyballs. Whatever shape you need to suit whatever cell part or enzyme you desire, chances are it is best draped on a carbon frame. Moreover again, the bond between two carbon atoms is one of the strongest bonds known, far stronger than that between two atoms of silicon, an element that otherwise has much in common with carbon. The strength of the carbon bond helps explain why it is the basis of life: we need molecular stability now, and we really needed it when life was new and the world was a considerably harsher place than it is today. At the same time, the carbon bond under ordinary conditions can bend, spring, and curl, hence the capacity of carbon molecules to array themselves as rings, cages, and coils. Carbon is as good as Goldilocks for building the spiraling, switchbacking molecule called DNA, and so the sugary spine of the double helix, and the individual chemical letters of which its code is composed, are carbonated through and through.

And while it may be mere coincidence, there's something gratifying about which gem we carbon vessels seize on prior to making a few carbon copies of our own: the diamond. Perhaps nothing underscores carbon's chemical genius better than the breadth of its packaging options, from the dark, slippery, shavable format of graphite on one extreme, to fossilized starlight on the other — translucent, mesmeric, intransigent

* Argon, krypton, xenon, and radon.

diamond, the hardest substance known, save for a human heart grown cold.

What spells the difference here, between carbon as ductile lubricant, a material you can spritz into balky locks, and carbon by De Beers? In graphite, each carbon atom is covalently bonded to three other carbon atoms, all of them lying in the same two-dimensional plane; there is no upstairs-downstairs blending of electrons but only the wan charms of van der Waals holding one floor to the next, so they slip-slide away.

In a diamond, by contrast, the bonds are fully fleshed out in every direction. Now, each carbon is strapped covalently to four of its kind, the maximum possible, and across three-dimensional space. To the left, right, crownward, groundward; wherever a carbon looks, there's a carbon bound to it. They're packed together so tightly and with such crystalline homogeneity that light finds very little impediment to its passage, very few imperfections to bounce off of and muddy the view, and the diamond gleams translucently. And because anywhere one might want to slice, one encounters thickets of jealous carbon-carbon bonds, a diamond feels like forever; to cut a diamond, a professional diamond cutter uses another diamond.

This painstaking compaction and positioning of carbon atoms is extremely difficult to accomplish. Getting every atom just where it needs to be to bond in a sororal three-dimensional mosaic, millions upon millions of flawlessly arrayed rings of four-faceted tetrahedrons, takes time and tremendous force. Until recently, the only place diamond factories could be found was hundreds of kilometers underground, in the Earth's mantle, where carbon stores subjected to great heat and pressure over millions or billions of years finally locked together in fixed constructs. Every so often, a volcanic eruption would spew a geyser of these diamonds to the surface, and another monarch might have his diadem, or Marilyn a pear-shaped friend. Industry also came to rely on diamonds for their unequaled ability to abrade metal machine parts into shape, and semiconductor manufacturers sought diamond bits to install in their microchips, to help prevent the embedded circuits from overheating. Diamonds happen to be excellent heat sinks, which is why even a room-temperature gem will feel cool to the touch. Put your fingertips or puckered lips against a diamond, and the jewel drains warmth from you to it, a heat transfer that your brain interprets as a brush with something cold; in fact, their high thermal conductivity, rather than their crystal clarity, earned diamonds the alias "ice."

Whatever the argot, diamonds clearly were too useful to leave to chance delivery through a magma pipeline. In the mid-twentieth century, scientists figured out how to mimic conditions in the bowels of the earth and began fabricating industrial-grade synthetic diamonds. More recently, researchers have managed to gin up gem-quality diamonds as well, although the process is so expensive that the resulting stone may cost a Tiffany customer more money than would a natural diamond from a Namibian mine.

Carbon bonds less zealous than those in diamonds help tape us together, and carbon bonds keep us alive. Most of the food that we eat, the carbohydrates, fats, proteins, and fiber, are carbon-centered compounds, which works out to an average daily intake of three hundred grams of pure carbon — about the weight of a pair of kidneys — per belly per day. Some of that ingested carbon is put to use directly, to repair damaged cells or assemble hormones, but more often the body simply cracks open the carbon bonds to extract the energy stored therein, and then tosses out the carbon atoms in the form of exhaled carbon dioxide. One species' waste is another's preferred taste, however. Plants blend the carbon dioxide together with water to synthesize sugars — and generate as a lucky byproduct the oxygen we need. The carbon cycle is just one of the many mandalas of life on which we rely, often mindlessly, and which we monkey with slapdashedly. Carbon bonds are strong, carbon bonds are dense packets of energy, and we can't get enough of them. The bulk of the energy driving the engines of our economy and our vehicles comes from sundering carbon bonds in coal, natural gas, and oil. Our cars, like our bodies, take the bond energy and dump the carbon, as carbon dioxide. Humans burn about 7 billion tons of fossil fuels a year, and so carbon deposits that might otherwise have hibernated underground for millennia instead combust into the atmosphere, harrying a carbon cycle that is already spinning as fast as it can.

The supple power of the molecular bond that gives us edible carbon fare and breathable oxygen pairs is crucial to life, but a covalent commitment can still be too ham-fisted when life demands Nijinsky. Here the secondary bonds come into play, and weakness becomes a source of strength. The backbone of DNA may be held together by carbon bonds, but the double helix is like a zipper, the "teeth" designed to fasten together or come apart as needed. If one of your body cells is about to replicate, for example, the two halves of the DNA molecule must separate to allow a copy, a carbon copy, to be made. If the cell isn't dividing, but simply needs to generate a fresh supply of an essential protein like hemoglobin or insulin, for example, the DNA must still zip apart, just a bit, to expose the spot where the recipe for hemoglobin is written in nucleic script. Open, shut, unwind, tighten up. The molecule of life jiggles viscously. Life was, after all, born in water, and DNA, the sentinel of heredity, doesn't forget its roots. The bond that holds the two halves of the helix together, that joins each tooth, or chemical letter, of one strand to the complementary tooth of the other strand, is the hydrogen bond, the same bond that keeps water clinging in the lugubrious bulge of a teardrop. The attractive tippiness that a molecule gains when a hydrogen atom shares electrons with a bigger, bolder, and more possessive element like oxygen, nitrogen, or carbon is just right for the needs of our genetic code. A hydrogen bond is strong enough to maintain the serpentine shapeliness of DNA during its tranquil hours within the nucleus, but it's an easy bond to interrupt for the sake of making new proteins or a whole new set of chromosomes.

The same goes for the protein molecules themselves. Proteins must have specific shapes to perform their tasks in the cell, but they must also be nimble, sinuous, squashable. Hydrogen bonds help define the most Gumbyish contours of a protein, allowing it to pleat out a bit on one side or buckle inward on another. Through hydrogen bonding, a hemoglobin protein can tangle itself until it looks like a plate of spaghetti and meatballs, each ball a lump of iron for clasping the oxygen we crave. Or an antibody protein of the immune system can array its four floppy chains into a straitjacket to form-fit any microbe encountered.

Sometimes, when you irreparably rupture a hydrogen bond, you can see life harden before your eyes. The clear liquid inside a freshly laid egg is a delicate matrix of some forty proteins designed to cosset a fetal chick through development, and those proteins owe their three-dimensional contours to hydrogen bonds. Fry the egg, and you destroy the bonds, freeing the protein components to rearrange themselves as haphazardly as they please. The buoyant, forward-looking, transparent syrup congeals into an opaque sedentary solid that now deserves the designation of egg white.

And while the hydrogen bond is first among secondaries when talking about the microscopic side of life, we organics mustn't neglect our van der Waals, the lumper and clumper and soft tissue wrangler. The layers of our visceral organs, and the figgy pudding furrows of our brain, are largely held together by van der Waals bonds. Our adipose depots in particular owe their cohesion to this loosest of glues, which is why it is easy to slice through fat with a steak knife or surgeon's scalpel — easier than breaking down the constituent fat molecules with exercise, for they are energy-rich stores of carbon bonds. Plants, too, rely on the van der Waals magnetism of their cellulose walls to survive. The internal surfaces of a plant's roots and stem are slightly charged, attracting water molecules from the soil and prompting them to begin crawling upward, as water will crawl up a paper towel if the corner of the paper is dipped in a puddle. Hydrogen bonds then ensure that more water molecules will follow their leaders along the cellulose road. The name of life is bonds, all bonds, an ecumenical band of bonds, each contributing what talent it can to uphold order, bolster morale, and resist the cosmic drift toward rot, at least for one more day.

Chemistry is about molecules and bonds, and it is also about a matchstick found, brandished, and at last ignited. "Chemistry is the science of change, the study of transformation," said Rick Danheiser. Its roots lie in alchemy, the ancient effort to turn lead into gold, the mundane into the glamorous, the dead into the reborn; the word "alchemy," from the Greek *khemia,* means "black sand," which the Greeks associated with ancient Egypt and its elaborate devotion to ensuring the pharaonic class a good life in the afterlife. The Chinese words for "chemistry" and "change" share a common ideogram, which shows a simple but unmistakable postural transformation, from a person standing to a person sitting.

The most unmistakable chemical transformation is that of a matter's state — a solid liquefies, a liquid evaporates, a vapor condenses into rain. For most of the furnishings of our everyday life, we associate a particular substance with only one of those three states. Wood, steel, and stone — solid. Oxygen and helium — gas. Alcoholic beverages — liquid (you can keep a bottle of Bombay Sapphire in the freezer, and somehow it remains an ever pourable starter to a gin and tonic). Water again bucks convention and seems almost equally at home in all three forms, as ice, steam, and liquid. In fact, Earth is exceptional in its possession of tristate water. Mars has a lot of water, but it's frozen away underground. Jupiter and Saturn have traces of water, too, but as orbiting ice crystals or as a gas among miasmic gases. Only on Earth are there ocean flows and Arctic floes and sputtering Yellowstone fumaroles; only the Goldilocks planet has water to suit every bear.

What accounts for the differences among a solid, a liquid, and a gas? And why are certain solids so reluctant to melt, while others begin to ooze out of the bag if you even think about taking them on a picnic? One obvious parameter you can fiddle with to induce a phase change in your sample is heat. Fry an ice cube, and the ice cube melts. By adding heat, you are amplifying molecular anxiety. Granted, molecules are fidgety from the start. Every bit of matter, no matter how sober its appearance, is constantly quivering at the base of itself; protons must spin, electrons gotta fly. In a solid, however — that is, a material with a fairly fixed shape and volume — the constituent molecules can only move so much, the strength of their individual motions counterbalanced by the firmness of their ties. As long as temperatures (and pressures) remain reasonably constant, the particles content themselves with Jack LaLanne isometrics and jogging in place.

Add heat to the solid, though, and the molecular oscillations quicken. The excited particles tug and strain against their bonds, and snap at their personal trainer, until a scattering of tiny rips appears in the three-dimensional array. Now the particles have some room to start sliding over one another. More sliding here means more gapping there, and more opportunity for the oscillating participants to shift out of their frame. When the last of the hindrances to intermolecular glide has disassociated, you're left with a liquid, a flowing substance that has a measurable volume but no fixed shape. If the liquid is heated more, the particles may gain enough kinetic dash to overcome any of the attractive forces that keep molecules clinging together and begin springing free of the surface, as a gas. The components of a gas still have their molecular integrity; the individual water molecules streaming out of your screaming teapot maintain their covalently bonded H_2O formulation; but they've shed all volume control and will diffuse out through whatever space you give them.

As a rule, ionic solids of rocks and bones are extremely resistant to melting and boiling. The rigid bonds holding ion to ion defy being loosened and pushed aside, the first step of liquefaction. Many a detective story converges on the telltale hearth, where the victim's skeletal remains refused to be stewed into silence. An ordinary wood fire burns at about 650° Fahrenheit and will barely make a dimple in teeth or bones; even the infernal 1800° of a professional crematorium needs two or three hours to dissolve the bulk of a decedent's skeleton, and still there can be bone fragments lingering in the ash. Metals, too, often have the mettle of Mephistopheles, and melt only at very high temperatures. Not only is the metallic bond that results from a sharing of electrons among multiple atoms quite strong, but the bartering system encourages metal atoms to pack themselves as densely as possible in three dimensions. The degree of solidity and resistance to melting varies considerably from metal to metal, though. Iron atoms have up to three electrons to share with their peers, and they stack together so closely that each atom touches twelve of its neighbors; iron won't melt until 2800° Fahrenheit, or 1538° Celsius. Herring-soft sodium, on the other hand, can share only one electron with its mates, and so a sodium-sodium federation is comparatively lax and will melt at 208°F. Silver, copper, and gold possess similar orbital architecture and all thaw down at somewhat less than 2000°F.

And then there is mercury, arguably the barmiest of all the elements. Mercury is liquid at room temperature, and it conducts heat and electricity so poorly that it barely merits inclusion in metaldom. Behind mercury's unusual behavior is its massive nucleus and the strong pull of its eighty protons. The positive packet at mercury's core keeps such a powerful lock on all the surrounding electrons that, even though the element theoretically has two negative particles to share in an electron sea, those electrons prefer staying close to their nuclear family, leaving the metallic bonds linking one mercury atom to another weak and easily disrupted.

Yet even as mercury's natal spirit is feckless and mercurial, the element readily forms soft amalgams with other metals, including silver and gold. The miners of ancient Egypt and Greece used mercury to extract gold from ore, and alchemists were convinced that if anything could transform lead into gold, it was the bobbling, quasi-animated metal they called chaotic water, or quicksilver. The magnificent Sir Isaac Newton, a passionate if episodic alchemist, considered mercury less a distinct element than a fundamental principle, the essence of all metals, and he sought it in its noblest and most "philosophical" form. Working in his Cambridge laboratory, Newton handled and sampled mercury droplets and inhaled their volatilized fumes, until he became as mad as a hatter or as flaky as a furrier — tradesmen that famously cured their fabrics in mercury and infamously suffered from the metal's neurotoxic effects. Preserved locks of Newton's hair reveal high concentrations of mercury, and, according to contemporary accounts, he grew increasingly hostile and choleric over time. Toward the end of his long life, the man who earlier had discovered the universal laws of gravity, motion, and optics and invented calculus, and whom James Gleick called the "chief architect of the modern world," expressed little interest in anything but that most fantastical of Gospels, the book of Revelation.

In contrast to ionic solids and the less mercurial metals, molecular solids are often disturbingly easy to melt and boil. This is especially true of solids that contain a mixture of different but closely related molecules, as do the soft organs of the body. Such solids are likely to lean heavily for their gross morphology on van der Waals, the promise most easily broken. A stick of butter, for example, which is about 80 percent fat and 20 percent protein, milk sugars, and other dairy components, melts at just about the temperature of the mouth — a concordance that in no small way explains butter's rich "mouth feel" and its inclusion in so many dishes we judge delicious.

Not every heated substance passes in orderly goosestep from solid to liquid to gas. Take frozen carbon dioxide, or dry ice, the basis of so many memorable children's birthday parties and forgettable stagings of *Macbeth.* On being exposed to room temperature, a block of dry ice bypasses the liquid stage altogether and evaporates directly into billowing white boas of smoke, an act of phase-change denial called sublimation. Dry ice owes its plumosity to both the relative frailty of the bonds binding carbon dioxide molecules together and the paucity of carbon dioxide in the lower atmosphere. At shirtsleeve conditions, the intermolecular links in dry ice quickly begin to dissolve, and the surrounding air essentially sucks the loosened rarities up wholesale and begs for more. Regular H_2O ice can also sublimate directly into vapor without pausing at the aqueous phase, though it does so much less dramatically. This is why ice cubes in a freezer tray gradually shrink despite the persistence of ambient frigidity. The circulating air skims off occasional water molecules from the top of the ice and eventually redeposits them as a scrim of frost on the sides of the freezer — or, if the cubes are loose in a container, as a kind of glue welding everything together into a Gaudíesque hoodoo of ice.

Melting, freezing, boiling, condensing, all represent physical changes in matter's state, but not in its composition. The molecular modules may get anarchic or they may get military, but they maintain their molecular identity. A rose petal is a rose petal, whether velvety-limp on the wedding room floor, or Popsicle-stiff in a liquid nitrogen bath. If you want something truly novel, you must change the substance chemically. You must break the extant molecules apart and reshuffle the subunits into new molecular configurations. If you want your loaves to leaven or your juice to ferment, neither boiling nor freezing nor squeezing will do. You need the pith of that allegoric black *khemia* on which the science of change is built. You need a chemical reaction. And what better way to summon the spirit of change than by raising a toast to the toadstool?

Fermentation may well be the oldest chemistry experiment in human history. Nobody knows how or when the first alcoholic beverage was made and sampled and declared "Satiny, supple, and exuberant, with notes of black fruit compote, sassafras, cocoa, cinnamon, meat, mineral, forest floor, Tigris, Euphrates, T'ang, and Tang.® Best if drunk before construction of the first ziggurats." Very likely the event was a total accident, the result of a few yeast spores blowing into a pot of mash that a careless child or a clinically depressed slave had forgotten to clear from the table. Whatever its origins, the vintner's art was domesticated soon after the advent of the agricultural revolution. Chemical traces on pottery shards from nine thousand years ago suggest that the citizens of Jiahu, a village in the Henan province of northern China, brewed a wine made from rice, grapes, and honey, a varietal that may explain why the best thing to drink with Chinese food is beer. And while alcohol has its desolate side, and has killed or made killers of millions, it also has kept millions alive. Through the many millennia before the advent of public sanitation measures, when water was notoriously nonpotable, people of all ages, at least in the West, often quenched their thirst with alcohol instead; given its mild antiseptic properties and its acidity, liquor was far less likely than water to carry parasites. The populace may have been slightly intoxicated much of the time, but better tipsy than typhoid.

Wine, beer, and other state-controlled spirits are the product of yeast cells feasting, and eating always requires chemical transformation: breaking apart molecules you find and using the parts and fuel to create the molecules you need. Yeast is a type of fungus, and while the fungal kingdom has an unusually catholic palate that may not always resonate with ours, the yeast strain that brews happens to share our love for sugar. If you add brewer's yeast cells to a vat of barley mash or well-stomped grapes, the yeast will latch onto the so-called simple sugars in the mix, "simple" meaning carbohydrate molecules that can't be broken down into still simpler carbohydrates. Simple sugars are the ones that taste sweet on the tongue and include glucose (which is the sugar that flows in your blood and serves as fuel to every cell), and fructose, the main sugar in fruit. (Put glucose and fructose together and you get sucrose, the table sugar you stir into coffee.) The two simpletons have the same chemical makeup, the same number of carbon, hydrogen, and oxygen atoms, differing only in how the atoms are arranged in three-dimensional space. No matter. The yeast will imbibe either, and will wrest energy from the sugar by breaking it down into two parts carbon dioxide and two parts ethyl alcohol, or ethanol. The carbon dioxide is the derivative that puts a little froth in the beverage, or, if the yeast has been added to bread dough rather than mash, that leavens its glutinous substrate into a puffy, oven-ready food item. The ethanol, of course, is what makes alcohol alcoholic, a leavener of mood and a lessener of sense. Ethanol is only one member in a large class of organic compounds called alcohols, colorless and flammable chemicals found in a wide variety of settings. With no help from yeast, your own body cells generate trace amounts of alcohol whenever they are forced to burn energy anaerobically, that is, without the benefit of oxygen, as they do during strenuous exercise like weightlifting, which is why a locker room can sometimes smell like a pub.

Regardless of source, all alcohols are accoutred with a hallmark hydroxyl group, a chemically reactive knob of oxygen and hydrogen that allows alcohol to wedge itself between comparatively bulkier molecules and help split them apart. Alcohol thus is widely used as a solvent in the manufacture of perfumes, dyes, pharmaceuticals, even children's cough syrup, and it makes for a pretty decent cleanser, too. Alcohol has low freezing and boiling points, allowing you to retrieve your designer liquor from the freezer and pour a neat shot right away, and to feel comfortable serving coq au vin to children or to Carrie Nation: by the time you take the pot off the stove, the alcohol in the wine sauce will have long since bubbled away.

Alcohol molecules can themselves be chemically transformed into sobriety. If you expose a bottle of wine to air and to the appropriate strain of aerobic bacteria — bacteria that need oxygen to feed and survive — the bacteria will pick up where yeast left off and break down the alcohol to water and acetic acid, or vinegar. As a molecule that dresses well with oil, vinegar has won its own measure of gastronomic fame at the salad bar; but for all its tart taste, vinegar lacks the inebriating vim of alcohol's hydroxyl accessory, and so could not addle a rabbit.

Fermentation is just a drop in the vast vat of reactive possibilities that surround us. Some chemical reactions occur easily and spontaneously while others won't bother unless you light a fire under their orbutts, or bury their starter parts underground and forget about them for a half a billion years. If you combine sodium and chlorine, poof, they'll react instantaneously, heatedly: Sodom meets Gomorrah, and we're left with a pillar of salt. And as the electrons of the participant ions assume their position in the crystal, they give up a bit of their verve, of their kinetic and potential energy. The total energy of the sodium chloride coalition is slightly less than that possessed by the sodiums and chlorides beforehand. Hence, the reaction that joins them is an exothermic one, a releaser of energy, in this case as heat, light, and the thrilling boom of a mini-explosion.

If, on the other hand, you stir together eggs, butter, flour, sugar, and other ingredients for a birthday cake, put the batter in a pan, and then realize halfway through the party that you never turned on the oven —

well, there's always Entenmann's. For the ingredients in the batter to react chemically and rearrange their bonds into the light, firm, moist, buoyant matrix of carbohydrates, fats, and proteins that we associate with cake requires energy. Baking a cake is an endothermic reaction, one that consumes heat rather than giving it out.

Then there are the chemical confrontations that start endothermically and end as a blast of hot air. The oxygen we breathe, the gas that makes up a fifth of our atmosphere, may be a lifegiver, but what a reactive zealot the molecule is. Oxygen combines with any substance it can, and in the merging steals electrons from its partner, changing the partner and singeing it and leaving it weaker than before. Oxygen is such a brilliant thief that the very act of electron piracy is called oxidation, even though other atoms and molecules can serve as oxidizers, too. Oxidation may be slow and steady, as it is when an iron bridge reacts with oxygen and starts to rust. Or it can be a matter of milliseconds: oxygen greets gasoline in the cylinder of a car engine, the mixture explodes, and you're on your way. Oxidative reactions are largely exothermic. A rusting bridge will emit modest amounts of heat, while the heat expelled by an internal combustion engine is great enough to warm a cat on a car hood hours after the engine has been cut. Combustion, though, generally requires an initial input of energy before it can turn self-sustainingly exothermic. A spark plug must spark the cylindrical courtship of oxygen and gasoline. A matchstick must be struck if it is to light in any way but figuratively. By scraping the match head on the appropriate surface, you heat it with friction. This heat is just what the sulfur, phosphorus, and other ingredients in the match need to combine in an unequivocally exothermic reaction. The heat from that sulfur-phosphorus collision in turn is enough to start oxidative combustion, the chemical confrontation between oxygen and a carbon-based substance — in this case, the wood shaft of the matchstick. Burning transforms the substrate into heat, light, carbon dioxide, and water vapor and will continue without further cajoling as long as there is carbon feed and oxygen greed.

Life is also a mix of endothermic and exothermic reactions, of gathering fuel and kindling, stacking the pieces with Boy Scout precision, striking the match, and feeling the burn. The body, after all, cannot afford to wait for the right chemistry to just happen. It doesn't have the luxury of sitting around for several million years like aluminum oxide, until the perfect confluence of geochemical events reinvents it as sapphire. Instead, the body must catalyze the reactions that it needs, pushing molecules together that might otherwise never find each other, and then bask in the energetic results of the chemical coupling. Our cells are replete with enzymes, proteins that make reactions happen in predictable fashion, just as the spark plugs of an engine keep the gas in the cylinders combusting. Digestive enzymes release the energy in food, liver enzymes detoxify poisons, immune system enzymes neutralize microbes. We take in fuel to generate our catalysts: enzyme fabrication is an endothermic enterprise. Many of those enzymes then catalyze exothermic reactions, keep tens of thousands, tens of millions, of tiny home fires burning each day, and in just the right way.

In life, as in love, timing is key, and even the wristless wear watches. Plants that enlist faunal mobility to the cause of floral ubiquity must be sure to maximize the sweetness and softness of their submissions just when their seed is set to be spread. They want you, the frugivore, to consume the fruit at that moment, slough off the packaging metabolically, and then amble away to void the indigestible seeds on some distant patch of maiden soil. The strategic ripening of an apple, then, tenders an excellent example of controlled carnal glee, the stepwise igniting of tiny chemical blasts that blaze up as color and fragrance and succulent roundness, all begging you to come have a bite.

Apples begin budding on a tree right after the blossoms of spring have enticed insect pollinators to help fertilize a new crop of seeds. The blossoms fall away, and, in a grand, endothermic production — paid for by the tree's photosynthesizing leaves — a fruit bulges up around five pockets, or carpels, of seeds. Those seeds need time to mature, however, before they are capable of leaving the pod and sprouting new apple trees. An unripe apple therefore is a forbidding fruit, its cell walls thick and impermeable, its meat starchy, fibrous, and acidic, its outer skin plasticine green — common fruit shorthand for CONSTRUCTION AREA: KEEP OUT.

Give the apple and its seeds time, however, and they begin releasing ripening hormones, most notably ethylene. Ethylene is a compact molecular bundle of hydrogen and carbon atoms — a hydrocarbon — but its effects are large and fruitful. As ethylene molecules diffuse through the apple in the manner of a gas, they stimulate the activity of other enzymes, a platoon of fruit gentrifiers, coaches, carpenters, copy editors, wardrobe consultants, attitude adjusters. Some enzymes clip the starchy, complex carbohydrates into simple sugars, others help neutralize the acids, while still others break down the pectin glue between fruit cells and so help soften the fruit. As the cells become looser, sweeter, and more permeable, the fruit adopts an almost animal-like respiratory style, breathing in oxygen and exhaling carbon dioxide. The soaring sugar content sucks in water from the stem, and the apple turns juicy. Its degraded molecules are now small enough to volatilize into the air and convey the distinctive aroma we perceive as apple. Enzymes in the skin help whisk away the green chlorophyll and generate in its stead bright, beguiling pigments of red or yellow, which can be seen from a distance and which are to a fruit-eating bird or mammal the visual equivalent of a dinner bell. Most of these chemical reactions are exothermic: in feel as in looks, the ripening fruit nearly glows. At last the apple can be plucked and sampled, and its warmth shared with someone you love.

6

Evolutionary Biology

The Theory of Every Body

As we were about to enter his office at the University of California's Museum of Vertebrate Zoology in Berkeley, Professor David Wake glanced off to the side and stopped abruptly.

"Wait a minute," he said. "I have to show you something. You'll love this. You'll absolutely love it." He darted over to a nearby shelf and retrieved from it a white plastic bucket with a lid on it. The lid had several holes punched through it. Professor Wake took off the lid and allowed me to peek.

"What the . . . ?" I sputtered in confusion as I stared into the bucket. At the bottom was some sort of extraordinary, lizard-shaped doll, but unlike anything I'd seen at a zoo gift shop, Toys "R" Us, or even the Blarney Stone cocktail lounge near Penn Station. Its five-inch-long body was light and shimmering, like semitransparent flan, and obviously molded from an advanced gel-solid polymer. Its head was tinted teal, its dainty legs and the tip of its nose bore a hint of Necco pink, and its back and fat tail were sprinkled with patches of copper and lilac. I couldn't stop gawking. Was it a replica of an ancient reptile, driven extinct by its insupportable distribution of pastels? Was it a kind of visual pun, created by an artistically gifted scientist as wry commentary on the entire field of herpetology? Was it for sale, or should I just steal it when Professor Wake wasn't looking? And, hey, how did he get the thing to blink and flick its tail just now without pushing any buttons?

"Isn't it the most beautiful creature you've ever seen?" Wake said. "It's a gecko. A colleague just brought it back from the Mideast."

"Wait a minute," I said, or maybe squeaked. "You mean this is a real, live gecko?"

"Live and in color," Wake confirmed. "It does have an unearthly and somewhat comical quality to it, doesn't it? Like something from Dr. Seuss. Or don't you think it would be a perfect model for the computer animators over there at, what's it called, Pixar Studios? They wouldn't have to change a thing." He snapped the lid back in place and returned the bucket to the shelf.

No, I thought to myself. The gecko is gorgeous. The gecko gets you from the get-go. But the gecko that I later learned was appropriately, colloquially, called the Wonder Lizard looks far too fake to make it in cartoons.

The fake fakery is part of the take-home message here. In biology, you should never believe your disbelief. There are so many species that arouse one's suspicions, that look too-too: too stagy, too silly, too gothic, too pastiched, too elegant, too composed, too momentous, too perfect. Every time I see a toucan, I'm dubious. Its hulking yellow bill seems out of all proportion to the rest of its body and just barely attached to its face, as though the bird had stuck its beak into a giant banana and decided it liked the effect. And speaking of improbable schnozzes, let's not snub the star-nosed mole, a semiaquatic mole found throughout eastern North America. Ringing its snout are twenty-two fleshy, pinkish red, highly sensitive tentacles that, when fully extruded and wriggling about in search of food, look like a pinwheel of earthworms, or children's fingers poking up from below in a cheap but surprisingly effective horror movie. Surely the star-nosed mole didn't just happen; surely there is a disgruntled employee in some dank basement cubicle to blame.

In fact, when nineteenth-century European naturalists first encountered the duck-billed platypus of Australia and New Zealand — with its shuffling, lizardlike gait, its beady little eyes and slits for ears, its webbed feet and oar-shaped tail, and that outlandish, rubbery, bluish black Marx Brother of a mouthpiece, which doesn't even have the courtesy to quack — they were convinced the animal was a hoax. Not until several platypuses had been killed and dissected were the skeptics placated.

Amazing grace can also look fake: Two trumpeter swans facing each other, heads bowed, foreheads touching, each balletic neck curved into one half of a heart. You watch them move, and you could swear they're aware of the power of their beauty, as though they live to make you wistful, humble, and in awe of the divine. Or a male painted bunting, red of rump and nape, blue of head, green of backside — a prince of primaries, a fistful of Matisse. I once saw a painted bunting on a log, and I couldn't believe how something so compact could fill my whole horizon.

Seated at his desk amid the biologist's customary organic habitat of rakishly piled printed matter and pantheistic bric-a-brac — a clock with frogs instead of numbers, a lively collection of genuinely fake reptile and amphibian statuettes, an old oven brick embossed with the word SALAMANDER, pictures of Charles Darwin, Ernst Haeckel, Richard Owen, and Homer Simpson — David Wake talked about his professional passions and his personal mission. He talked about tree frogs, skinks and sticklebacks, salamanders and their slingshot tongues. He described his atypical biography, a familial emulsion of two mindsets, the theological and the scientific, which normally carp at each other like Montagues and Capulets, red states and blue; and he talked of how his hybrid background informs the rigorous zeal of his teaching. Wake reminded me of a Methodist minister I knew in grade school, the father of one of my best friends — the same whitening hair and bespectacled, gentle blue eyes, the same open, loving-kindness kind of charm. But where Mr. Hill was an evangelist for revelation and the Gospel, David Wake prefers evidence and a really good fossil.

"I was raised in a conservative Christian community," Wake told me. "My grandfather was a Lutheran pastor, my parents were very religious. I myself went to Pacific Lutheran College. Two of my cousins have doctorates in theology. One served as the president of a Lutheran college in Alberta, and another was a bishop in Canada. So, you see, my family is filled with religious people and theologians.

"At the same time, my family also has a lot of scientists to its credit. One cousin was a curator at the Field Museum in Chicago. Another relative is a curator of the Natural History Museum in Oslo. My grandfather, the pastor, was an amateur naturalist. He lived in our home for a while, and he lived to the age of ninety-nine, so I knew him well. And he never, in his long, rich life, felt any conflict between his religion and his scientific knowledge. Nobody in my family did. My grandfather was the one who first taught me about evolution. He taught me to respect evidence and to remember that religion must always accommodate reality. We live in the real world, he said, and we must understand the world on its own empirical terms."

Wake has a message to share, and it is one that virtually all the scientists I spoke with, no matter their field, ranked at or near the top of their list of things they wish the public understood about science. The message is the alpha and omega, lox and bagels, of the life sciences.

Theodosius Dobzhansky, the great Russian geneticist, said it pithiest: "Nothing in biology makes sense, except in the light of evolution."

Evolution. *Evolution.* EVOLUTION! It doesn't matter whether you're an atheist, a churchgoer, a craven Faust in a foxhole. You may be Catholic, Muslim, Hindu, Jewish, Druid, a born-again Baptist, a born-again-and-again Buddhist. It doesn't matter what you believe to be our purpose here on Earth or hope to find in the hereafter, or whether you have faith in a Supreme Being or prefer the Ronettes. It doesn't matter what disk you insert in the mental module marked "God." None of it will suffer if you see the principle underlying and interlocking all earthly life. The life that we see around us, the life that we call our own, evolved from previous life forms, and they in turn descended from ancestral species before them. Newer species evolved from prior species through the majestic might of natural selection, a force so nearly omnipotent in its scope and skill that it needs no qualification, supplementation, ballast, or apologist. Evolution by natural selection, which also goes by the name of Darwinian evolution or Darwinism (darwinism on "casual Fridays"), explains life on Earth in its outrageous entirety, all the 30 million or 100 million species here today — many that have yet to be counted and classified, let alone inspire the next blockbuster tie-in — and all the many hundreds of millions of creatures that have arisen and vanished in the several billion years since life first appeared. For many biologists, evolution is part of the definition of life. "What is life?" one researcher put it. "That which eats, that which breeds, that which is squishy, and that which evolves."

Darwinism is so essential to understanding the slightest attogram of biomass that even physicists agree it should be granted equal protection in the eyes of the law. "People like to think of physics as being the source of the fundamental laws of science," said the MIT physicist Robert Jaffe. "But there's one fundamental law that comes from the life sciences, and it's just as deep and all-pervasive and universal as anything in the pantheon of physics. Evolution by natural selection is an absolute principle of nature, it operates everywhere, and it is astonishing. But evolution is underappreciated, and, what hurts me far more, it is under assault."

Darwinism is by no means universally despised or rejected. To the contrary, evolutionary theory has a rather large fan base, and as David Denby wrote in *The New Yorker* several years ago, evolutionary biology has replaced Freudianism at dinner parties as a preferred source of speculation for why this or that friend is behaving so badly. Charles Darwin's distinctive profile, the long white beard, the Victorian frock coat, may be second only to Einstein's as a scientific face recognizable to a good chunk of the lay public. In many parts of Europe, Asia, and Latin America, evolutionary science is a staple of science education, and has no more currency as a source of sociocultural angst and spitting than might Copernican ideas about heliocentricity. Nevertheless, in America, home to many of the greatest research universities in the world and to more Nobel laureates of science than any other nation, the battle against evolution madly, militantly, proptosically, soldiers on. It may be wearing a moth-eaten uniform and carrying a musket, and its side may have lost the evidentiary war more than a century ago, but drat it all to hell, the gun still shoots: a guerrilla war against the monkey huggers!

Again and again, the opponents of evolution have managed to keep evolution from being taught in our schools or have demanded that biology textbooks also present "alternative viewpoints" to evolutionary theory, including nonscientific, data-deprived ideologies like creationism and intelligent design. The campaign against Darwinism has been successful enough to plant kernels of doubt in many minds. In one recent poll, which echoed survey results from the last couple of decades, only 35 percent of American adults agreed with the statement that "evolution is a scientific theory well supported by the evidence." As years of education mounted, so, too, did support for Darwin: 52 percent of college graduates and 65 percent of those with postgraduate training expressed acceptance of the theory of evolution. Still, that leaves 35 percent of the most scholastically saturated Americans looking askance at one of the bedrock concepts of modern science.

I'm always surprised at how often I encounter resistance toward or doubts about Darwinism among otherwise rational people. When I was thinking of writing a children's book about evolution, for example, and I asked my cousin, an artist, if she might illustrate it for me, she said she would — even though she didn't really believe the whole ape-to-human storyline. Another time, while I was standing around talking to a perfectly pleasant couple at a friend's wedding near Sacramento — he a lawyer, she a businesswoman — I mentioned evolution as a jumping-off point to another subject I had in mind. My conversation partners stopped me right there. "So," said the lawyer, "I take it this means you have no doubt that evolution is for real?"

"Um," I replied, staring into the crystal depths of my champagne glass, which was, tragically, empty at the moment. "About as much doubt as I have that, if I were to let go of this glass, gravity would pull it to the floor, it would shatter to pieces, and the bride would be pretty upset because it's a Waterford."

The couple chuckled weakly and then realized that a dear friend on the other side of the room was either calling their names or should be.

So maybe I don't get invited to many parties. Nevertheless, among scientists, the matter is as settled and straightforward as I made it sound. You release your glass, it falls to the ground. You gaze out at nature, evolution all around.

"The evidence for evolution?" said Tim White, a paleoanthropologist at the University of California, Berkeley. "Overwhelming and incontrovertible."

David Wake, who for thirty years has taught a course at Berkeley in advanced evolution, said, "The evidence is rock-solid, firm, and unassailable." Whenever you take medicine, he pointed out, chances are it was first tested on laboratory animals before being approved for use in humans. You may believe the Earth to be only six thousand years old and every creature installed as is by the Lord thereon; yet still you'd feel a tad safer knowing your sacrificial guinea pig had been a rodent rather than, say, a spider or a snail. "Why do experiments on mice more than on spiders, if not for the reason, as we all innately understand it, that mice are more like us than spiders are?" said Wake. "Hmm. Could that have something to do with evolution?"

Richard Dawkins, an evolutionary scientist at Oxford University, indefatigable defender of Darwinism, and author of *The Selfish Gene, The Blind Watchmaker,* and other lucid splendors, made the eloquent evolutionary case yet again, in an interview with a reporter from Salon.com. "It's often said that because evolution happened in the past, and we didn't see it happen, there is no direct evidence for it," he said. "That, of course, is nonsense. It's rather like a detective coming on the scene of a crime, obviously after the crime has been committed, and working out what must have happened by looking at the clues that remain. In the story of evolution, the clues are a billionfold."

There are clues from the distribution of genes throughout the animal and plant kingdoms, he said, and from detailed comparative analyses of a broad sweep of physical and biochemical characteristics. "The distribution of species on islands and continents throughout the world is exactly what you'd expect if evolution was a fact," he continued. "The distribution of fossils in space and in time are exactly what you would expect if evolution was a fact. There are millions of facts all pointing in the same direction, and no facts pointing in the wrong direction. The British scientist J.B.S. Haldane, when asked what would constitute evidence against evolution, famously said, 'Fossil rabbits in the Precambrian.' They've never been found. Nothing like that has ever been found. Evolution could be disproved by such facts. But all the fossils that have been found are in the right place."

You can't pull Bugs Bunny from a billion-year-old hat, and pterodactyls never tugged on Raquel Welch's thong. "You have to be diabolically blinded," said Wake, "not to see evolution in everything that we do."

A good part of the problem stems from one little word: "theory." That Darwinism is called "the *theory* of evolution by natural selection" invites popular confusion and leaves the science vulnerable to determined adversaries. Will you look at that? say the critics. Scientists themselves call evolution a theory, rather than a fact. Obviously they must have doubts. And if they have so many doubts, why shouldn't the rest of us? For that matter, why should we believe their theory, their "creation myth," rather than somebody else's? As a bumper sticker I saw recently put it: THE THEORY OF EVOLUTION: A FAIRY TALE FOR GROWNUPS. In some states, antievolutionists have demanded that stickers be put on high school biology textbooks to point out that evolution is "just a theory," not a "fact."

Fie, fie to scientists here, for using a word like "theory," which has the common connotation of "conjecture," "speculation," or "guess." A pretty good guess, maybe even an educated guess, but still, a theory is a "could be" and not a "proven fact." Normally, I'm no promoter of technical jargon, but in this case, I wish scientists had a word of their own, to mean what a theory means to them. A solid, pompous, unflinchingly scientific term, in the style of "ribosome" or "igneous." A phrase resistant to casual or calculated misapprehension and to the juggernaut of justs.

Sometimes a cigar is just a cigar, and a trip to the moon on gossamer wings can easily be just one of those things. But a scientific theory is never just a Just So story. In science, an idea that has yet to be put to the test or burnished by evidence is called a hypothesis. You notice something about the world, and you propose a possible mechanism to explain the observation. There's your hypothesis. The hypothesis could be the result of simple reasoning by analogy, an extension of previous findings onto your similar though not identical case study; or it could be sheer heliospheric speculation. However sensible or sensational the conjecture may be, it's not your theory, it's your hypothesis. It's *just* your hypothesis. To test the hypothesis, you design an experiment, or you gather a generous sample of data points from the field, and you become a freak for control groups. You analyze your results and needle them with statistics. Now you have a result. If the result vindicates your initial hypothesis, go ahead and crow. If not, go ahead and dream up a new and improved hypothesis to explain retroactively the findings you found; that's what the discussion section of a scientific paper is for. Either way, the verifiable, irrefutable fact remains, you don't yet have a theory to bear your name.

A scientific theory, like Einstein's theory of general relativity, like the theory of plate tectonics, like Darwin's theory of evolution, is a coherent set of principles or statements that explains a large set of observations or findings. Those constituent findings are the product of scientific research and experimentation; those findings, in other words, already have been verified, often many times over, and are as close to being "facts" as science cares to characterize them. To take a simple example: entomologists are always discovering previously unknown species of insects. You hike in the Adirondacks, they say, you poke around in the Great Lawn of Central Park, and you, too, might unearth a new type of beetle, which you can then offer to name after the police officer threatening to fine you for defacing city property. There are tens of millions of insects waiting to be discovered, of a staggering medley of sizes, disguises, noises, and knacks. Yet for all the diversity, entomologists know that any new insect they stumble on will display the following characteristics: three body segments — a head, thorax, and abdomen; three sets of legs; and a hard outer shell, or exoskeleton. These facts of insecthood are so robustly established that they're part of a Spanish song my daughter learned in kindergarten: "*¡Soy insecto, a veces pequeñito! ¡Seis piernas para caminar, cabeza, tórax, abdomen, abdomen, abdomen!*" The traits are the shared, taxonomically defining characteristics of the insect class, and they are the result of all insects having descended from a common ancestor. Here, then, is a modest factlet, one among legions, that is best regarded and understood beneath the vast climax canopy, the grand explanatory framework called the theory of evolution. Why do so many of earth's creatures have six legs, three body bays, and a stiff outer coat? Because the 30 million or so insect species alive today descended from an ancestral specimen bearing that winning combination, a real *pequeñito* of a progenitor that lived some time during the Devonian period, around 400 million years ago. But why do crickets, dung beetles, dragonflies, head lice, hornets, termites, praying mantises, and the rest of the teeming clade look so different from one another? It's descent with modification. As the insects radiated outward and began inhabiting a range of niches, they evolved to suit their station. Natural selection stepped in, brandishing a fly swatter, and, Whoa, it sure is a good thing I happen, through random mutational change, to resemble the leaf I'm sitting on.

Either fact you focus on, the diversity among insects or the traits that bind them, makes sense only in the light of evolution. The *theory* of evolution.

Or compare the following foursome of forelimbs — a bat's wing, a penguin's flipper, a lizard's leg, a human's arm. On the surface, the various appendages look quite dissimilar from one another, and they perform distinct tasks: flying, swimming, darting, purchasing hands-free electronic devices. Yet beneath the miscellany lies skeletal homogeneity, for each forelimb houses the same set of four bones: humerus, radius, ulna, and carpal. The bones are splayed in the bat's wing, converged to a V-tip in the penguin's flipper, but they are readily seen on an X-ray as anatomical homonyms. Embryonic development further confirms the link. If you were to watch a time-lapsed video of how the respective fetuses grow — the baby lizard and penguin in their eggs, the bat and human in their wombs — you would see the four forelimb bones budding out from the same prefatory parts in the embryo. Such structural and developmental cronyism tags us all as descendants of the first vertebrate tetrapods, the valiant four-legged forebears that traded the seas for the soil. Their basic skeletal structure proved so fit for the challenges of terrestriality that all vertebrate forelimbs are modified meditations on the humerus-radius-ulna-carpal theme; we wear the tetrapod coat of arms up our sleeve.

Haldane's droll observation about the proper placement of petrified rabbits is another firmly grounded finding, a fact that must be faced. You won't find rabbit fossils in a bed of trilobite remains. Trilobites, those familiar paleopetroglyphs that look variously like horseshoe crabs, cockroaches, and Game Boy video icons, were for 300 million years the dominant life form in Earth's oceans. There were more than 10,000 species of trilobite, ranging in size from a millimeter — the comma you just passed — to creatures as long as your arm. Trilobites were bottom feeders, seaweed grazers, trilobite biters. They breathed through and swam with their gills, and they had eyes like no other. Whereas the lens of a standard eye is constructed of protein molecules, the trilobite lens was like a marble chip, made of the mineral calcite. Yet by a quarter of a billion years ago, those sharp little eyes had had it. Trilobites went extinct at the end of the Permian period, along with 90 percent of all other marine species then on Earth. The last trilobite likely imprinted its image in perpetuity about 20 million years before any remains of the earliest mammals can be spied in the fossil record, let alone of the first rabbit, which dates from a mere 57 million years or so ago. Paleontologists have seen it and shown it again and again. Fossils are found in the right place and from the right time, newer fossils stacked in layers above older fossils. Trilobites abounded around the world, and whether you're digging in Australia, Austria, or Cincinnati, their fossils are always located in sedimentary beds at just the depth and relative position you'd expect them to be for a creature that thrived a half billion years ago. The same for dinosaur fossils, or the bones of all the archaic, outrageous mammals from the Oligocene, circa 35 million years B.C., like *Indricotherium*, or "giraffe rhinoceros," the most massive land mammal ever and a decent brontosaurus manqué; *Archaeotherium*, a boarlike beast the size of a bull and with scythes where you'd think slicing canines would do; and *Cainotherium*, a distant hoofed relation of the camel but with the face, ears, and forelegs of Haldane's mascot, the rabbit. Wherever you find fossils, they fit. Giraffe-rhino fossils are found in strata that can be dated to the Oligocene, and those fossils are stacked above the Jurassics and the Triassics. The consistency and sequential structure of the fossil record are all facts, big fat faceable facts, and they've been backed up more often than the freeways in L.A.

This is a scientific "theory": not a hunch, not even a bunch of hunches, but a grand synthesis that gathers "facts" or robust findings with petty p values, and infuses them with meaning. A scientific theory also has predictive power. Under its rubric and tutelage, you can generate new ideas about how the world works, and then put those ideas to the test. By using evolutionary reasoning, for example, you might come to certain conclusions about the relationships among different organisms. Long before scientists understood anything about our genetic code, the swizzler stick of DNA that thinks it calls the shots, they had classified organisms into kinship cliques based on their anatomy and behavior. They determined that mice and humans were mammals, and that mice had much more in common with us than did spiders: their organs, brain, cardiovascular system, chemical composition, immune system, reproductive habits, limb and eye count, all were closer to ours than were those of a spider or a fly. Hence, geneticists could easily predict that the murine-human relatedness would extend right down to the threads of our genes, to the individual bases, the chemical subunits, of which our DNA is composed. Sure enough, as scientists began spelling out the genetic codes of a variety of organisms, they found that the closer a creature was to a human macroscopically, the closer it was alphabetically. Mouse DNA is about 70 percent identical to ours, while that of a fruit fly is 47 percent.

Scientists could go further in their predictions about molecular genealogy. Sure, we're genetically closer to mice than to flies, but how can it be that nearly half of our DNA still resembles the recipe for a creature with compound eyes, backward-bending legs, and a persistent desire to take up residence in a Porta-John? For that matter, we share one-fifth of our genetic code with yeast, an organism that has neither *cabeza, abdomen,* nor anything else requiring multicellularity. What sorts of genes could possibly be tying us to fungus?

As it happens, there are many basic chores that every cell must know how to do. Whether of wildebeest, baker's yeast, human humerus, or fly glomerulus, a cell must be able to take in nutrients, throw out the trash, stay in shape, and divide when told. One would predict, then, that the genes encrypting such fundamental tasks would be the genes least likely to change over evolutionary time, no matter who inherits them — and that is precisely what geneticists have found. The cell's maintenance and division genes are among the most well preserved specimens nature has to offer. We should all look, after half a billion years of evolutionary heaving and hawing, as dewily unchanged as do the genetic instructions that tell a cell to split in two. In fact, science has put the timelessness of DNA's blue-collar codes to spectacular use. Through studying the genes in yeast that oversee cell division, researchers have learned far more about human cancer than the malignancies themselves would deign to divulge. Tumor cells are ugly, messy, hard to handle. Yeast cells are pliable and generous (remember, they gave us wine). Should we ever declare victory in the ragged "war on cancer," the theory of evolution can claim credit for having sharpened the spears.

Another reason evolutionary theory may sometimes seem less bedrock-solid than it is stems from some of the internecine haggling among evolutionary biologists over details — the sort of squabbles that for almost any other scientific discipline would be of interest only to the contestants and their listservs; but with Darwinism as the national blood sport, everybody wants to be cc'd. Evolutionary biologists do argue over the mechanics of evolutionary change, how fast it happens, how to measure the rate of evolutionary change, whether transformations occur gradually and cumulatively, putter and futz, generation after generation, always working to stay ahead by a nose, until, whaddya know, you're wearing a Chiquita on your beak; or whether long banks of time will pass with nothing much happening, most species maintaining themselves in a comfortable stasis until a crisis strikes — an asteroid hits the Earth, or volcanoes dress the skies in flannel pajamas of sulfur and ash — at which point massive evolutionary changes may arise very quickly.* They debate what constitutes a species, and where you draw the line between two truly distinct species, each worthy of formal codification through Linnaean nominalization, and two different subpopulations of the same species. They scuffle over the nexus between evolution and romance, and whether a female chooses to mate with a male based on her careful assessment of his underlying genetic quality; or because she noticed that every other female was chasing after the guy and figured maybe they knew something she didn't; or because his nose reminded her of a favorite food item, and she was hungry.

Yet no matter how they swat the details, evolutionary scientists do not dispute the fundamentals. They do not argue over the reality of evolution, or that existing species evolved from previous species. And they do not dispute the engine that drives evolutionary change, as elucidated so brilliantly by Charles Darwin and Alfred Wallace 150 years ago: natural selection. Natural selection is the force that transforms drift and randomness into the gift of extravagance. It takes the doctrinaire sloth of the second law of thermodynamics, the tendency of every system to get frowzier over time, and hammers it into a magic, all-purpose, purpose-making machine that turns around and breaks entropy at the knees.

The basic premise of natural selection is simple. Parents give birth to multiple offspring — more offspring, as a rule, than can be expected to survive. Those offspring are like their parents, but not exactly like them. Each child's DNA is a uniquely shuffled, braided, clipped, and restated version of its elders' DNA. Genes that had lain dormant in the mother find their spine in the young, while a dominant trait in Dad is silenced come the son. And then, there may be a few total novelties in the mix. A new mutation, a slight change in the chemical spelling of a gene as it is bequeathed from parent to progeny: What do you expect, when you're copying out DNA, a sentence 3 billion letters long? Life, like everybody else, makes mistakes, and mutations are part of the fun. As Yogi Berra pricelessly put it, "If the world were perfect, it wouldn't be." And if there weren't thermodynamically inevitable bugs in the DNA copying program, we'd all still be bugs of a different sort:

* The late Stephen Jay Gould favored the latter scenario, which he called punctuated equilibrium: evolutionary stability as the norm, punctuated every now and then by mass extinctions and evolutionary overhauls; and because Gould, as a scientist, essayist, and best-selling author, did more than anybody else to introduce Darwinism to a popular audience, his take on the subject accordingly has been widely aired.

unicellular, genetically identical archaeobacteria happily burbling by a hot spring, as the Old Ones did in the beginning.

Through mutations and DNA shuffling, discrepancies arise in the gene pool that give nature something to select from. Now nature has choices among the plethora of offspring; let the winnowing begin. A microbe is born with a metabolic mutation that allows it to digest more and grow bigger than can its compatriots on the stromatolite program. The microbe greedily consumes whatever resources it can wrap its fatty acid membrane around, including — hey! — the poor mother bacterium that happened to spawn it. Soon the hyperphagic youth begins spawning spores of its own, many outfitted with the advantageous metabolic mutation, and the tribe drives the more sedate unicells into oblivion. A few hundred or hundred thousand generations go by, and another happy gaffe arises, this time in a gene that dictates the performance of a component of the microbe's membrane. As a result, the microbe proves unusually sensitive to cues from its neighbors, able to tell who's where, what they're doing, and how to profit from their labor. Before you or anybody else knows it, this ancestral bugging device has bred an army of eavesdroppers, and the world of unicellular, non-cooperative, asocial narcissism gives way to multicellular, interactive, community-based narcissism. In the wake of this sublime innovation, feudalism, monarchism, democracy, plutocracy, the postmodern corporation, Monopoly, and Clue were sure to follow.

Natural selection, then, is a two-step exercise of almost unlimited potential. First, minor inherited variations arise in a population by chance. A frog is born with a mutation that lends her head an odd, rhomboid cast. Other frogs stare and make rude belching noises as she hops past. The croak, of course, is wholly on them: little Braque girl turns out to look just enough like a fallen leaf to blend into the forest floor when an amphibivorous bird comes pecking, and so she outlives her taunters. Additional mutations among her descendants fortuitously enhance the camouflage effect, and every time a better cloaking device arises, natural selection favors the bearer just enough that the mutation soon becomes the species norm. Today, the renowned Solomon Island leaf frog looks so much like a leaf that, again, you can't help but shake your head in near disbelief. Ridiculous! How can a random mutation just "happen" to carve a few corners into a frog's figure? How can a deviation of an ordinary amphibian gene whip up something so goofy that also happens to be so useful? Let alone a string of random genetic changes that just happen to improve on the masking effects of the mutations preceding them. What are the odds that a series of snafus would shake out as the perfect disguise?

Quite high, in fact. Frogs are under relentless pressure from a broad range of predators. Birds, snakes, turtles, mammals, other frogs, scorpions, tarantulas, Jacques Pépin — all seek the dense, crunchy energy packets that frogs and their legs embody. A few industrious vipers can wipe out more than a hundred frogs in a single twilight hunt.

But frogs compensate for their extreme vulnerability by breeding like a certain other prey species that hops. Beyond ensuring that at least some frogs will survive to reproductive age, fecundity breeds evolutionary opportunity. Given the great number of froglets produced in a single generation, the episodic appearance of wonderful blunders is to be expected; and every defensive leap forward will be quickly selected. Soon the accidental has become foundational, the species standard from which other revisions may or may not arise.

Insects, too, have both the incentive and the mechanism to evolve a dazzling pageant of "Where's Waldo?" routines. Everybody in the world eats insects, either willingly or between visits from the city health inspector. Insects blunt the sting of brevity with stunning fertility. Among my favorite examples of prolificacy is *Blatella germanica,* the common German cockroach. If left unchecked, a single female can, in her twelve or so months of life, give rise to 40 million offspring. With insect odds, all bets may as well be FDIC-guaranteed. The dun coloration of the German roach suits its urban habitat, but whatever the occasion, an insect can dress for it. The Javanese leaf insect not only sports the central rib and radiating veins of a leaf, but also the appearance of little holes and torn edges that you'd expect of a leaf partly eaten by . . . insects. A stick insect looks like a stick and acts like a stick, which means not too stuck-up or suspiciously static. Just as a real tree part twists in the wind, so a seated stick insect will intermittently, woodenly, sway back and forth — animal imitating vegetal imitating dryad at night. But my vote for the dandiest drop-dead disguise goes to the swallowtail caterpillar, which resembles freshly deposited guano.

In the vast clan of insects and their arthropodal relations, all conceivable strategies have been sampled, all weapons amassed. Imitation, obfuscation, threat of death or indigestion — name your poison, there's an arthropod bartender ready to serve. The whip scorpion sprays a scorching one-two cocktail punch: oily caprylic acid to penetrate even the toughest outer sheath of an attacker, and water-based acetic acid to burn the tender tissues beneath. The devil's rider walkingstick backs up its defensive camouflage with chemical artillery, shooting streams of terpenes, potent chemicals similar to the active ingredient that makes catnip detestable to everything except, inexplicably, cats; and to see a blue jay get hit in the face with walkingstick spray is to see a blue jay that will never again question a twig. Some millipedes contain high doses of a progesterone-like compound that may serve as a long-term defense strategy by crimping the fertility of millipede foes. That outcome would not be of much use to the consumed specimen serving as the sacrificial Pill, but, in reducing the number of future predators, it would give a leg up to the millipede's surviving relations.

Insects have the means and motive to synthesize more defensive chemicals than we humans have had time to tally or test. They also have a flair for outwitting us when we turn our chemical arms against them. When we think of the dismal history of DDT, we think of springs silenced of birdsong and skies brushed free of bald eagles. Yet the real failure of our insecticidal campaign against mosquitoes and the diseases they carry was in how quickly the buzzing suckers came to shrug off our sprays. By the time DDT was banned in the United States, in 1972, nineteen species of mosquitoes — about a third of the known malarial vectors — were immune to the pesticide. Have your dependent loved ones perchance encountered any head lice lately? If not, then either you are a homeschooler, or your kids are very unpopular. In recent years, *Pediculus capitis,* a bloodsucking parasite with a particular fondness for the comparatively soft scalps of children, has joined the schoolyard metal detector and the thirty-pound backpack as a staple of modern childhood. The reason is simple: head lice have become murderously hard to kill. They're virtually immune to soft-core toxins like pyrethrins — the active ingredients in lice shampoos sold over the counter — and pediatricians are understandably reluctant to recommend that stronger poisons be applied within seepage range of impressionable young brains. That leaves tedious and inefficient parental nitpicking as the primary defense against lice, which pretty much guarantees that there will always be a parasite reservoir somewhere, ready to infest fresh heads in Topeka today or reclaim seasoned ones in Des Moines mañana.

The pace at which insects become resistant to our poisons, and bacteria to antibiotics, is often fast enough to observe. One year, baited traps took care of our ant problem; the next year, I watched in horror as the ants marched right through the little hockey puck disks without missing a ta-rah, en route to their main course at the cat food dish. Or the crickets that populate and freely defecate in our basement: Every spring, my husband sprinkles poison in all their favorite crannies and hatcheries. Up until last spring, the treatment worked, and the crickets crumbled. This year, either the treatment didn't work, or it worked the way radiation did for ants in the 1950s sci-fi classic *Them!* If you have doubts that evolution happens, I invite you to stop by our basement, where the crickets now look like kangaroos.

Whatever the pest in question, the evolution of pesticide resistance conforms to the Darwinian algorithm. Random genetic variations arise all the time in a population, especially among fast breeders. Most of those variations are either of little consequence or are decidedly inadvisable, and they are accordingly ignored by selective pressures, or are quickly swept from the gene pool. Every so often, however, a mutation of enormous utility like toxin resistance springs up, and the novel trait soon becomes the species norm.

Genetic quirkery can also sow biodiversity. If a mutation happens to affect a key gene that controls an animal's basic development, the resulting aesthetic or behavioral changes may be so profound that the beneficiary of the mutation looks or acts like a whole new species. And if that remastered organism and its progeny somehow become separated from their unmutated peers — say, by rising sea levels that transform a peninsula into an island — the odd stock may indeed evolve into a distinct species that will no longer couple with its former fellows. Scientists at Stanford University, for example, recently traced the evolution of the stickleback fish family to a handful of comparatively simple genetic changes. There are about fifty species of stickleback fish found throughout the Northern Hemisphere. Some live in the ocean and are protected against predators by a full-body armor of thirty-five bony plates. Others swim through freshwater lakes and rivers, freed of their cousins' cumbersome chain mail and thus able to dart about speedily and feed competitively. The researchers have determined that a few mutations in a single gene underlie this dramatic discrepancy in stickleback anatomy, specifying maximum plate growth in the marine fish, more or less suppressing it among lake dwellers. The ease with which major overhauls in stickleback format can be accomplished helps explain why the family managed to diversify its ranks so rapidly: freshwater sticklebacks diverged from their oceanic counterparts a mere 10,000 or so years ago, at the end of the last Ice Age, and they have further speciated and specialized themselves in whatever body of water they have colonized.

The evidence for evolution abounds, within us, beneath us, crowning and surrounding us. Antievolutionists complain about the "gaps" in the fossil record, and lacunae there are, by the chasm. Several hundred thousand fossil species have been identified and named, but researchers suspect that the known bones represent a mere one-thousandth of one percent of all species that have lived. "Of course there are plenty of gaps in the fossil record," said Dawkins. "There's nothing wrong with that. Why shouldn't there be? We're lucky to have fossils at all." Think of all the obstacles that a corpse must overleap en route to immortality. First, it must avoid the fate that awaits most dead organisms: getting picked apart and scattered by scavengers; decomposed by worms, mold, and microbes; husked and battered by the elements; or any or all of the above. The best defense against nature's multilateral recycling program is a quick burial, which generally requires dying someplace where a thick layer of sediment is likely to sweep over the carcass soon after it is deposited: in or around lakes, rivers, swamps, and lagoons, for example, or on the ocean floor close to shore and its sandy, silty runoff. The sedimentary blanket helps to prevent decomposition, at least of the organism's toughest tissues — bones, teeth, shells, tusks, woody stems. Over time, the sand and silt sedimentation turns to stone, and so, too, may the bones and other bioremains within, as, bit by bit, mineral particles come to replace the original organic molecules while maintaining their positional integrity.

Yet even then, a fossil is not safe. Its sedimentary cemetery may end up getting buried under so many subsequent layers of rock that bed and fossils are melted beyond recognition, or are ripped apart by the constant shuffling of the crust's plates, or blasted to ash in a volcanic snit. Finally, there is the considerable problem of discovery. After spending tens of thousands to millions of years patiently petrifying underground, a fossil must fight its way to the surface again if its record is to be read. It must hitch a ride on the edge of an uplifting plate, to emerge on the exposed side of a hill or mountain. Or it must be on a sedimentary plateau that has been painstakingly carved into by wind or water, revealing a Dagwood-sandwich stack of rockmeats, a fossil feast. Paleontologists do most of their hunting on hillsides and canyon gullies, where fortuitous collusions of geology and meteorology have served up samples of archaic sediments, glimpses of long ago that, on most of the earth, are stashed far below.

The ease with which paleontologists can affix an age to an outcrop of rock and the fossils embedded therein varies from site to site, but it can nearly always be done. Fossils younger than 55,000 years or so can be dated with high accuracy by measuring the ratio of two forms of carbon, carbon 14 and carbon 12, that still linger in the organism's remains; the less carbon 14 there is relative to carbon 12, the older the fossil. When it comes to appraising older fossils, scientists must look to the stone that houses the bone. Rocks as they form often become infused with a host of so-called radioactive isotopes, unstable versions of atomic elements like uranium 238, potassium 40, and rubidium 87 that attempt to tranquilize themselves by periodically and methodically spitting out excess particles from their unwieldy cores. Because scientists have determined the rates at which these radioactive atoms decay into stability — a pace known as the element's half-life — they can use the changeling tracers as geological clocks. Their Geiger counters clacking, researchers compare the proportion of still volatile to safely spayed isotopes in their disinterred treasure, and so they can get a reasonable grip on how long the rock and its entrenched fossils had been festering underground, to time frames dating back hundreds of millions of years.

Yes, it's hard to be a fossil, harder still to be a found fossil with a dependable isotopic birth certificate. Of course there are gaping gaps in a record so reliant on a defiance of nature's tireless composting piety, and on the blind luck of revealing uplifts a million years hence. Wouldn't you be wary if the gaps were too *few?*

Besides, gaps are perpetually being at least partially plugged. In 2001, researchers digging in the low hills of northern Pakistan, at a site once submerged beneath the warm, shallow waters of the Tethys Sea, discovered two superb caches of whale fossils that help trace the mammal's brazen backward flip from terra firma to aqua primordia. Biologists had long assumed that whales — which in their streamlined design and fealty to water seem so piscine that Herman Melville deemed them fish, but which breathe air, breastfeed, and bear hair follicles like any other mammal — were descendants of land mammals that returned to the oceans some 50 million years ago. The whale's fossil trail, though, was so spotty that biologists could only guess at what the prenautical whale might have looked like. Now they have compelling evidence that the progenitor to Moby Dick looked and ran like a wolf but ate like a pig, for it was closely related to ancient artiodactyls, the group of hoofed ungulates that today includes pigs, camels, cows, and hippos. Even before venturing into the water, the new fossil trove shows, the protocetacean had specialized ear bones that now are found only in whales and dolphins, suggesting that the whale's impressive audio skills, its capacity to hear freed Willy keening half an ocean away, may have evolved to track sound prints on land. Instead, it heeded the songs of the sirens and followed them into the sea.

Moreover, there are some beautiful fossil series that show persuasive procession of one species into the next. Among the most famously fleshed-out fossil record is that for the horse. According to the sequence disinterred to date, the first horselike genus was *Eohippus*, or *Hyracotherium*, an agile, four-toed creature the size of a Labrador retriever, which daintily dined on shoots, berries, and leaves in the woodlands of Eocene North America, 53 million years ago. *Eohippus* gave rise to several lineages, horses of different sizes, toe numbers, teeth ridges, and, undoubtedly, colors, although fur does not fossilize well so we'll probably never know. Some species were suited for life in the deep woods, others for open grasslands. One successful savanna specialist, *Hipparion*, migrated across the Bering land bridge to the Old World about 10 million years ago, and soon spread across southern Eurasia and Africa. Back in North America, all of *Eohippus*'s descendants gradually died off, and by the start of the Pliocene, 5 million years ago, only the horse called *Dinohippus* remained. Large and rugged, *Dinohippus* had long legs and single-toed, padless hooves ideal for galloping across the open plains that proliferated as the climate grew cooler and drier. It also had big, thickly enameled teeth designed for a lifetime of grinding on tough scrub grass and the tougher silica that inevitably comes up with it. *Dinohippus* begat a slightly more graceful version of itself, *Equus*, the modern genus of horse. At some point, *Equus* crossed paths with the comparably toothsome and leggy *Hipparion*, and for whatever reason — greater fecundity, lucky horseshoes — managed to supplant it. All of today's breeds of horse, from a grizzled Central Park carriage horse to a thoroughbred, sparrow-boned stallion, are members of the equine club. So, too, are existing species of zebra and wild ass. Their evolutionary heritage is a canter cast in stone.

Another group for which the fossil record is surprisingly rich is . . . our own. If you want to implicate a supreme being in the story of human evolution, you might consider inserting it here, as the mastermind behind the fortuitous events that yielded a wealth of prehuman remains in settings that normally are hostile to fossils. A supremely levelheaded being who wants only to guarantee that those in a position to ponder their roots have a look at the family tree. We have fossils of primogenitor primates from 80 million years ago, shrewlike mammals that began spending less time on the ground and more in bushes and trees, where they evolved large, forward-facing eyes well suited for ferreting out insects and dexterous fingers for plucking found insects off leaves. We have fossils from 50 million years ago, when the archaic arborealists started to diversify and give rise to early monkeys and apes, and these fossils formed in the semitropical forests of Africa, where humidity and armies of mulchers generally reclaim all biodetritus before it has a chance to be archived in sediment. We have souvenirs from ancestral apes like *Dendropithecus, Proconsul, Kenyapithecus.* "There actually seem to have been more potential ancestors than we would have needed prior to 12 million years ago," the Stanford biologist Paul Ehrlich writes in *Human Natures.* We have, he added, "an embarrassment of fossil riches."

So, too, do we have a sturdy chain of "missing links," of fossils with a variegated mix of traits we might call either "humanlike" or "apelike." There's lovely Lucy, the petite australopithecine female — named after the Beatles song "Lucy in the Sky with Diamonds," which was playing in the tent the day Donald Johanson discovered her — who clearly stood upright but whose skull was a quarter the size of ours; and such early hominids as the semibrainy *Homo habilis,* presumed to be one of the first users of stone tools, and the horse-toothed *Homo ergaster,* and the lantern-jawed *Homo rudolfensis,* and *Homo erectus,* skull swept back as though in a shower cap, and archaic *Homo sapiens,* and early modern *Homo sapiens,* and fully modern *Homo sapiens,* and that quintessential caveman, *Homo neanderthalensis,* which some researchers call Neanderthal while others prefer to drop the *h,* but all agree has been unfairly maligned in serving as shorthand for "extremely primitive, unhygienic, liable to grunt." Neanderthals coexisted with *Homo sapiens* throughout Europe for at least 100,000 years before dying out suddenly, even catastrophically, about 28,000 years ago. The reasons for the Neanderthals' demise remain unclear. Their brains were as large overall as those of their *H. sapiens* peers, although their skulls were of a slightly different shape, flatter and more beetle-browed, suggesting that they had a comparatively smaller frontal lobe, the part of the cerebrum we prize as the seat of our intelligence. Neanderthals, like *Homo sapiens,* fashioned fine stone tools with sharp, flaked edges, but they seemed less interested in art and ornamentation, in painting cave walls or carving pieces of ivory into female figurines with imprudent body mass indexes. Some skeletal evidence suggests that Neanderthals were far more prone to injury, arthritis, and other debilitating conditions than were *H. sapiens.* Or maybe our ancestors couldn't stand the sight of those lowbrows, and exterminated them. Genetic studies strongly suggest that there was never any romantic intermingling of the two hominid species, no evidence that we carry traces of Neanderthal genes. Whatever the cause, when the Neanderthals departed the scene, only one member of the *Homo* genus remained, we the self-designated drivers and namer of names, with our high, proud foreheads and our three-pound brains. *Homo sapiens sapiens,* so wise we had to say it twise. How can a sentient *sapiens* look at a lineup of hominid and prehominid skulls in a natural history museum and not be impressed at the traits that bind us, and those that set us apart? Descent from a common ancestor, modification by natural selection.

The fossil record, as sketchy as it is in spots, is unerringly consistent in sequence, no matter where in the world you dig. The molecular record, too, reveals the relatedness of all living beings, and also corresponds as one predicts to the evolutionary branching patterns of the major organismic tribes. Our genes are much more similar to those of a mouse than to those of a fly, and they are more similar still to the genes of a chimpanzee, our closest living relation. If you take one strand of a double helix from a human cell, and line it up against a helical strand from a chimpanzee cell, the two strands will stick together — will find the chemical counterpart they expect — along all but 2 to 4 percent of their spans. The 3 billion or so chemical letters that make up our DNA are 96 percent identical to those of a chimpanzee. Looked at another way, 120 million little genetic bases, just about enough to fill one of the 23 sausage-shaped chromosomes you see if you examine your fetus's DNA through amniocentesis, is all that partitions tourist from tenant at the San Diego Zoo. Which is what you'd expect for two species that shared a common ancestor only 5 million years ago. A mere 250,000 generations back, a quarter of a million great-great-greats; a forebear so far, yet so near, I can't help but call it Grandpa Silas.

Another big chunk of evidence for the theory of evolution can be seen in the realm of biogeography, the distinctive distribution of species around the world — at least until we humans started redistributing species willy-nilly as we wandered. Darwin was deeply impressed by the spatial clustering of what he called "closely allied" organisms — species with similar body plans and characteristics. Latin America, for example, is home to a magnificent, highly endangered, and inexplicably obscure family of birds called cracids (rhymes with "acids"), fifty species of large, meaty creatures notable for their vivid variety of headgear: the piping guan's foppish Mohawk crest, the helmeted curassow's bright pink knob bulging up like blown bubblegum between its eyes, the long blue horn of the appropriately named *Pauxi unicornis.* Cracids can be found as far north as the Texas-Mexico border and as far south as Buenos Aires Province — although overhunting and habitat hacking have drastically reduced their numbers — but because they don't fly well, they haven't crossed any oceans. You won't spot a wattled guan in the

Laotian rainforest or a *Mitu mitu mitu* sunbathing on the Solomon Islands. Neither would a wild penguin be caught dead in the jaws of a polar bear. All eighteen species of penguin live in the Southern Hemisphere, many of them around Antarctica, while the polar bear, like its close cousin the grizzly bear, is strictly a resident of the north. In the eyes of biologists from Darwin onward, the concordance between geography and biology, the clustering of "closely allied" species on the same landmasses and the discrepancies between the inhabitants of one continent and those of another, can be traced back to one elegant explanatory engine. "We see in these facts some deep organic bond, prevailing throughout space and time," Darwin wrote. "This bond, on my theory, is simple inheritance." The descendants of a common ancestor, sharing common ground.

Yet another line of proof is best captured with a line of mnemonic ditty: "Kings pour coffee on fairy god-sisters." This is my favorite way of remembering the taxonomic system that we use to classify species. You have your kingdom, then your phylum, your class, order, family, genus, and species. It's a nested sequence of categories, from big-picture suzerain to a specific little sister — the word "specific," conveniently enough, being the etymological progeny of "species." The narrower the niche, the more traits the pigeonholed will share; the broader the category, the larger the number and the wilder the heterogeneity of its members. No matter how hurly-burly any corraled crowd becomes, though, the beings bunched together in one batch will have more in common with one another than they do with those in any other like-tiered grouping. Let's take a quick look at ourselves. We *sapiens* are the only living species in our genus, *Homo,* although the fossil record shows there have been other *Homos,* like Neanderthal and Erectus, before us. Our family is Hominidae, and we share it with four living species of great apes — chimpanzees, bonobos, gorillas, and orangutans — as well as dozens of extinct predecessors of varying apely or humanesque traits. We hominid apes join with some 200 species of monkeys, lemurs, tarsiers, lorises, and the like in the order Primate; and with another 4,600 or so species in the class Mammalia, a cadre united by our hair, four-part hearts, two-part ears, and motherly udders; even those egg-laying outliers of mammaldom, the duck-billed platypus and the anteating echidna, dribble milk from their mammary glands that their hatchlings lap up. Our phylum, Chordata, subphylum Vertebrata, celebrates our backbone, and brackets us together with more than 50,000 other vertebrates, like reptiles, birds, fish, and amphibians. Our kingdom is Animalia, and here we run into the great throngs of arthropods and other spineless animals: insects, spiders, scorpions, millipedes, lobsters, crayfish, and crabs; and oysters, octopuses, gastropodal makers of dye; and the worms and the sponges, corals, sea pens, sea cucumbers; many millions of animae with wide-open mouths or mouth pores, defined by our need to feed on somebody, somehow. Not so for the 260,000 species in the kingdom Plantae, those hidebound Rumpelstiltskins that spin sun into gold; yes, even the Venus flytrap, should no insect come calling, can rally its chlorophyll and get by eating light.

If we continue clambering up the tree of life, however, into a fairly recent addition to the classification scheme that has yet to be incorporated mnemonically, we'll join with trees and other plants, as well as with algae and yeast. Above kingdoms are two "empires," the eukaryotes and prokaryotes — we eukaryotes being those whose cells are equipped with a nucleus in which the double helix is cradled, and prokaryotes, like bacteria, with their DNA floating unbounded in the viscous cell belly, the better to divide if you just give it twenty minutes. And if you rise higher still to view the code by which life carries on, eukaryote and prokaryote become one. Inside every cell, and every viral parasite of a cell, you will find the same chemical alphabet, the same nucleic acid letters that tell a single epic story in a billion different ways. Above empire, kings, and caffeine, we have the Gaia of genes.

As the writer and naturalist David Quammen has observed, this phylogenetic sorting system, this nesting of category within category, and the tiered pattern of resemblances that brings ever more species into the fold and finally culminates in a single ancestral supertrait — the shared chemistry of our genes — is not the ordinary way we organize collections of items. I, for instance, have a large collection of bookmarks from around the world, dating back to the early nineteenth century. I've organized the antique ones by theme — the bookmarks that advertise pianos, or Pear's soap, or chocolates, or tires, or Smokey the Bear, or the Scottish Widows Fund — but there's no systematic way to link the tire bookmarks to the perfume bookmarks to the Mr. Peanut 1939 World's Fair commemorative bookmarks. The same goes for my daughter's collection of boxes. She likes to arrange them in aesthetically pleasing configurations, but there's no obvious morphological hierarchy, no reason for saying the jeweled box is more like the painted wood box than it is like the carved wood box. Why can't bookmarks and boxes, or rocks, or earrings, be systematized like matryoshka dolls? Because, Quammen writes, "Rock types and styles of jewelry don't reflect unbroken descent from common ancestors. Biological diversity does." And the number of traits two species share, or the degree to which their

DNA strands might happily, stickily, intertwine, is often a measure of how recently the two species diverged from a common ancestor.

Yet not every case of similitude in nature is proof of a close bloodline. Sometimes organisms on one continent will bear a startling resemblance to species located halfway across the globe, to which they are only very distantly related. For example, the cacti of the Americas are quite difficult to distinguish from a group of African succulent plants called euphorbias. In both families, you have some species that are shaped like slightly squashed dough balls and others that grow tall and upright, like aspiring totem poles. Euphorbias and cacti display a similar preference for spines or thorns over leaves; are sheathed with a thick, waxy skin; and store water in their hollow cores. If you bought a euphorbia and nicknamed it Saguaro, your aunt from Tucson might not see any cause to correct you. Yet the cactus and euphorbia families are as scantily related as two plant groups can be, and each has much nearer floral cousins without a spine to their frame.

The same with the anteating echidna of Australia, the anteating pangolin of Africa, and the giant anteater of Latin America. The three mammals share more than a fondness for ants and termites. Each has an extenuated, depilated snout, wormlike tongue, bulging salivary glands, a stomach as rugged as a cement mixer, vestigial teeth, and little scythes on its feet. Yet the trio's last common ancestor probably darted among dinosaurs. The echidna, remember, is still laying eggs, and its nearest kin, the platypus, looks like a Muppet.

Importantly, the anteating trio, the bicontinental succulents, and a plethora of other cases in which the anatomy matches but the taxonomy clashes only serve to underscore Darwin's sweeping authority. All exemplify the phenomenon of convergent evolution, of widely dispersed lineages confronting similar problems, and, through the guiding hand and cracking cat-o'-nine whip of natural selection, independently devising the same basic solution, the same set of tools to get the taxing job done. Both the euphorbias of sub-Saharan Africa and the cacti of the Sonoran plateau of North America have evolved in some of the harshest, parchest, and most sun-beaten habitats on earth, and there are only so many ways for a plant to weather a life *in extremis*. You can adopt a round conformation, which lends you the least amount of surface area relative to your volume: that way, you have a minimum of covering exposed to the harsh sun and drying winds, but a relatively big central holding tank to store whatever water may fall during a brief desert shower. Alternatively, you can grow tall and upright, so that little of your surface sits directly beneath the glare of the midday sun, while again giving you internal space for a personal reservoir. Leaves increase your total surface area and wick away moisture from within, so best to dispense with them entirely and let your stem do the photosynthesizing. Thick waxy skin inhibits evaporation and deters the sharp incisors of thirsty desert rodents, while thorns not only add to the defense against hydrotheft, but also help channel dewdrops and rainwater down to the plant's shallow roots. Yes, if you plan to succeed under fire, you'd better have a tough hide and plenty of big pricklers on your side.

Another risky profession is myrmecophagy, the consumption of ants, and it doesn't help if you plan on ordering a side of termites. Ants and termites are among the most successful of all arthropods, such a dominant presence in whatever habitat they choose to colonize that other insects like beetles or cockroaches are consigned to puddling around their outskirts. Edward O. Wilson has estimated that ants alone make up at least half of the world's insect biomass. Much of the success of ants and termites lies in their social skills, their ability to work together seamlessly as highly specialized but de-individualized members of their collective — to behave as a ruthless "superorganism" and model for the Red Menace of McCarthyism and the spandexed Borg of *Star Trek*. Nowhere is the insects' militant nationalism more evident than in their commitment to homeland security. When attacked, ants and termites reply en masse, stinging, biting, shooting out streams of formic acid, swarming into eyes, ears, nostrils, pants. Hence, while an ant colony or termite mound of millions of individuals would seem to present an irresistible target to nearly every passing food pipe, in fact for many creatures resistance is prudent. If you have designs on this refractory form of sustenance, you can't be an amateur or do it part time. A hammer won't work; you need specialized gear.

In taking up the challenge to exploit a vigorously fortified resource, the echidna, pangolin, and giant anteater have converged on the same safecracking utensils: large, sickle-shaped claws for digging into nests; a long, sticky ribbon of a tongue for poking deep into the dugouts and lapping up hundreds of insects per probe; an elongated muzzle for precision firing of the tongue; a denuded muzzle so that ants and termites have no fur to grab on to for a counterattack; enlarged salivary glands to keep the tongue gummy and to help wash the ants down; and an ironclad stomach to withstand all the sand that accompanies every ant sampler. Horses have big teeth to endure the silica contaminants in grass roots. But with no need to chew their tiny prey before swallowing, anteaters opted to preempt dental angst and have forgone tooth eruption completely. Kent Redford, a biologist with the Wildlife Conservation

Society in the Bronx who has studied anteating animals, admits that theirs is a "weird bioplan," but one with box office legs. When you see multiple lineages independently evolving a similar morphology, he said, you've got to figure the recurrent design is the obvious choice, the most natural selection.

Convergence, camouflage, Donald Duckbill and Toucan Scam. Wherever you rummage through the emperor's phyletic cabinet of curiosities, you'll see how nonrandom, how purposeful, Darwinian evolution can seem. If much of nature looks designed, that's because it *is* designed. Not from the outside in, but from the inside throughout, on the fly, by life striving to fulfill the prophecy of itself, and to remain, at all costs and by any pathway or laugh track, here on Earth, among itself, alive. Critics of evolution complain that a purely Darwinian or "mechanistic" explanation of life consigns us to a life stripped of meaning, to a world driven by random forces, exigencies, and pointless amoralities. Gregg Easterbrook, a writer who has been described as a "liberal Christian," has posited that "the ultimate argument will be between people who believe in something larger than themselves," that is, those of deep religious faith, "and people who believe it's all an accident of chemistry." Yet this binary formulation is needlessly inflammatory and far too penurious. What is the "all" that is to be explained as "an accident of chemistry"? The all of biological diversity with which the world overbrims? To characterize life as embodying accident is highly misleading, no matter what your spiritual leanings may be. Life is the anti-accident, the most thermodynamically profligate heave-ho ever instigated and subsequently amplified, annotated, explicated, expurgated, renovated . . . well, you get the idea. We don't know how life began, but even that first replication of an unknown molecule was not really an accident. It was lucky that the conditions were right for the replication to occur, perhaps, but the very act of self-copying was, in its way, a deliberative act. The inherent tautology of the definition of life — that which lives and seeks to perpetuate itself — already removes accident from the equation. Indeed, there are some origin-of-life researchers who insist that, under certain conditions, life is virtually inevitable. Are these conditions rare enough to qualify as genuine accidents of chemistry? Or do they abound throughout the universe, a consequence of hydrogen and oxygen being among the commonest of all elements, and therefore water, the fount of life, being one of the commonest of molecules? We don't know yet, but I can say that the great majority of astrophysicists are convinced that we earthlings are far from alone, a subject I'll return to in the book's final chapter.

However incidental or inevitable were life's beginnings, its efflorescence into the "all" we see around us has not been at all accidental or random. "Natural selection is about as nonrandom a force as you can imagine," said Richard Dawkins. This is not to say that natural selection has specific goals in mind, or that it has proceeded in stately forward march to engineer progressively more complex and intelligent organisms, an effort of which we, of course, are the crème flambé. Natural selection seeks only to select that life which knows best how to live, and sometimes, as the archmodernist Adolf Loos said, ornament is crime. For example, the tunicate, or sea squirt, is a mobile hunter in its larval stage and thus has a little brain to help it find prey. But on reaching maturity and attaching itself permanently to a safe niche from which it can filter-feed on whatever passes by, the sea squirt jettisons the brain it no longer requires. "Brains are great consumers of energy," writes Peter Atkins, a professor of chemistry at Oxford University, "and it is a good idea to get rid of your brain when you discover you have no further need of it."

Evolution is neither organized nor farsighted, and you wouldn't want to put it in charge of planning your company's annual board meeting, or even your kid's birthday party at Chuck E. Cheese. As biologists like to point out, evolution is a tinkerer, an ad-hocker, and a jury-rigger. It works with what it has on hand, not with what it has in mind. Some of its inventions prove elegant, while in others you can see the seams and dried glue. "The assumption often is that organisms are optimal," said Bob Full, a materials scientist at the University of California, Berkeley. "They are not. Organisms carry with them the baggage of their history, and natural selection is constrained to work with the preexisting materials inherited from an ancestor. Dolphins have not reevolved gills, no titanium has been found in tortoise shells, and you would never design a bat from scratch."

Why are posters of the Heimlich maneuver hung in every restaurant, and why is it so easy to choke on a pretzel? The evolution of human language was made possible by our larynx dropping down from its previous primate position, thereby opening up a larger air space to facilitate elaborate sound production. In addition, the position of the tongue changed. Whereas a chimpanzee's tongue is contained entirely within the mouth, the back of a human tongue forms the upper edge of the vocal tract, giving it flexibility in shaping and articulating sounds. Those twin modifications incidentally brought our food and air pathways much closer together than they had been in our prehuman ancestors, or than they are in our latter-day ape kin, with a concomitant rise in the risk of an embarrassing and potentially fatal episode of a bite of pickle dropping into the trachea rather than the esophagus, where it belongs. By themselves, the laryngeal mutations would have been swiftly whisked from the gene pool, but throw in the novel capacity for orating, educating, obfuscating, browbeating, backstabbing, filibustering, and singing jingles in the shower for products you don't even like, and *now* you're talking.

Moreover, not every feature on a creature is the product of natural selection. Some may be residual traits that are no longer needed or functional, but that are harmless and so are not under selective pressure to be tossed off like so much tunicate cerebrum. When we are cold or alarmed, for example, we get goose bumps, which may look cute on children emerging from a pool but which don't do them nearly the good they might if the children still had fur. The reaction harks back to our pelted past, when the raising of serious body hair helped to lock in heat during the cold, or made one look bigger in confronting a foe. Other traits arise in one sex not for direct utilitarian purposes, but because they're crucial in the other sex, and the basic body plan of mammalian embryonic development happens to be bisexual. Witness the male mammal's compact, dairy-free nipples, generally the same number on him as will be found on his lactationally competent counterpart — two on a man, a male chimpanzee, a male bat; ten on a male dog and eight on a male cat.

A still greater engine of pomp, camp, and comedy, of conspicuous traits that may do nothing to increase an individual's life span and in some cases help clip it short, is the evolutionary force called sexual selection. Darwin himself described this impressive complement to natural selection and offered extensive evidence of how the need to attract a mate and thwart one's rivals can have a radical impact on an animal's profile and behavior. Even traits that seem to impede a species' capacity to escape from a predator or to fade securely into the background — the standard bequests of natural selection — will find evolutionary favor, Darwin said, if they so enhance their bearer's sexual appeal that they end up swamping the competition in the gene pool. After all, if you survive long enough to breed, and if you score handsomely, even orgiastically, in a single spring spree, who cares if you're a feather duster come summer? Your seed will do your future struttings for you. The classic illustration of sexual selection at work is the peacock's tail. A peahen is a dowdy, mostly beige bit of bird-dom, but she clearly has lurid appetites. Over many generations of peahens preferring males with ostentatious posterior plumage, peacocks have evolved tails so cumbersome that they can scarcely flap up to the lowermost branches of a tree — presumably a potential handicap for a bird native to the land of the leopard, a famously agile climber. Nobody knows why peahens like the male tails they do. Is it something about the depth and purity of the iridescent colors that signal the male's underlying genetic worthiness? Or do the peahens attend more closely to the quantity and symmetry of dark eyespots on the plumage, spots made especially prominent against a shimmering emerald and turquoise backdrop? Could it be the capacity and willingness of the male to carry the weight and to fan it wide whenever they pass that the females find so fetching? Whatever message the tail conveys, it is one no proud peacock can afford to forgo.

The ferocious struggle not only to attract a mate but to fend off rival suitors can also leave an evolutionary cross hair to bear. Each mating season, male deer do little beyond butting their racks together, until finally the lesser racks roll, and the triumphant stacks become studs. The annual intermale joustings have placed a high premium on the possession of large, sturdy antlers that can take a beating without cracking and on multiple forking prongs for latching into a rival's rack and flipping him over. The elaboration of the stag headdress has accordingly mounted over time — and is sometimes mounted over a human hunter's mantelpiece, too. Conversely, the males of many spider species are tiny, a fraction of the size of the female. She has pressing need of her heft: to catch prey, to spin silk, to lay eggs. The he-spider has need only of speed, to reach a receptive mate in advance of all those other eight-legged wallets of sperm. But the male's slight size leaves him defenseless against the female if she's in the mood for a postnuptial snack. As ever, love hurts.

William Saletan once wryly observed in the online magazine *Slate* that evolution doubters, like any other group of organisms, can be organized taxonomically. The ancestral members of the lineage are the straight-up creationists, those who interpret Genesis literally, believe the Earth to be only 6,000 years old, and insist that all species, including people, were built by God as is, *in toto* — and that goes for Dorothy and her little dog, too. No Darwinism, no natural selection, no *Eohippus* meets *Dinohippus* and talks to Mr. Ed, no cockamamie picky peahens, no Permian period. No evolution, period.

This founding credo of hard-core creationism has been around for many, many decades — it was the stimulus for the famed Scopes monkey trial of 1925 — and it shows no signs of going extinct. In recent years, biblical literalists have managed to persuade the U.S. Park Service gift shop at the Grand Canyon to carry their coffee table book, *Grand Canyon: A Different View,* which alternates gorgeous sunset photographs with the argument that the canyon is the handiwork of Noah's flood. A lavish new Museum of Earth History in Eureka Springs, Arkansas, displays fastidiously detailed models cast from genuine fossils of *Tyrannosaurus, Thescelosaurus,* and other dinosaurs, but positions them side by side with Adam and Eve and attributes the dinosaurs' demise largely to that explanatory Zelig, the Flood.

Nevertheless, the pressures of contending with the mountains and arroyos of evidence that attest to Earth's great antiquity, and to the evolution of species through the eons, have resulted in their own speciation event. Strict creationism has given rise to new species, new attempts to undercut the reach of Darwin's theory of evolution by natural selection. Perhaps the most notorious of the derivations is intelligent design, and though the phrase is meant as a tip of the hat to a presumably divine designer, it never hurts, if you're going to start a fight, to claim the word "intelligent" for your side.

Creationists, as a rule, reject the evolutionary account of when species arose and how to understand the great galloping biological diversity on which human health and all future Sierra Club calendars depend. Creationists see nothing implausible about a museum diorama depicting dinosaurs grazing alongside woolly mammoths because they are unpersuaded by the stratification of the fossil record that would separate the animals by at least 60 million years — a figure that in any event exceeds their estimate of Earth's age by a factor of 10,000.

Advocates for the idea of intelligent design, on the other hand, are quite willing to accept the geological evidence that our planet is about 4.5 billion years old, and they concur with the mainstream view of a biological timeline that dates back several billion years. They have no quarrel with the proposition that humans arose from apelike progenitors, nor with the general capacity of whole organisms to change over time and give rise to new species. A number of the ID heavyweights are scientists, most vocally Michael J. Behe, a professor of biological sciences at Lehigh University in Bethlehem, Pennsylvania. "Intelligent design proponents," he wrote in an op-ed piece for the *New York Times,* "do not doubt that evolution occurred."

Where the ID ideologists part company with the preponderance of scientists is on the origin of the smallest components of life — our cells and the enzyme and protein "machines" that keep our cells and selves thumping. As Behe and his sympathizers see it, cells and their microcircuitry are almost too good, too well composed, too perfect, to be believed. Many of the protein complexes essential to life, they say, work only if all the parts are present and pulling their weight. Should one participant in these molecular assemblages fail, should one spring go sproing, then the entire structure collapses. In other words, when you go below the gross scale of the body, below the squishy, messy organs of the body, and get down to the fundamental units of a body, you start to encounter elegance, beauty, something they call "irreducible complexity." The protein partnerships that run the show could not have arisen gradually, they say, through random mutations and modifications of preexisting structures. The molecular components of the cell are too interdependent, too carefully arrayed, to be the product of ordinary Darwinian natural selection. Natural selection requires that intermediary stages in an evolving structure lend an advantage to their recipient over the structures that preceded them. If you're a frog that looks a tiny bit like a leaf, you'll have a tiny survival advantage over another frog that looks purely frog, and so a leafy camouflage can evolve in stepwise fashion. A slight widening of the skin on the front legs that allows an arboreal mammal to get a bit of lift as it leaps from a branch may help it escape predation, and therefore you can imagine the gradual triceps extension that leads to the winged bat. But with these molecular assemblages, the IDers insist, there's no in between. The pieces must all be in place, watches synchronized, or you get system failure. Natural selection doesn't work on complex, interdependent assemblages, they say. If draft versions of a product flop completely, they won't be selected, and you'll still be stuck up a tree.

Among the examples that advocates often cite to illustrate the irreducible complexity of life's foundational widgetry are the tiny hairlike cilia by which paramecia and bacteria propel themselves through the water; the daisy-chain pathway of proteins that conveys light signals from the eye to the brain; and the intricate blood-clotting mechanism that keeps us from hemorrhaging uncontrollably with every nick of an envelope edge. In each case, multiple proteins cohere into unity and act with patriotic fealty, one nation, indivisible. If you destroy any one of the sixty or so proteins of which a paramecium's cilium is constructed, the hair doesn't beat more weakly or slowly than before. It cannot beat at all, and the protozoan isn't goin'. The clotting response that comes to our rescue when we fumble a morning shave is a tightly choreographed cascade of ten distinct protein "factors." If only one of those factors is defused by an inherited genetic mutation, you can end up with hemophilia, or "bleeder's disease," whereby the slightest injury can kill you. How can something as complex and essential to survival as the clotting reaction possibly have evolved through clunky, mincing Darwinian mechanisms, Behe wonders, and the gradual plunking of one Lego piece onto another, when a defect in just a single step brings the entire business to a standstill — or, in the case of blood, to *not* stand still?

If the fundamental modules of the cell and of our biochemistry are irreducibly complex, Behe continues, and if they cannot be explained as the fruit of conventional evolutionary forces as we understand them, and if in fact they look to be the miniature set pieces of an immortal genius, an infinitely scientifically notated Leonardo, why rule out the possibility that . . . they just might be? Why not leave room, at the very base of life, for the contributions of an intelligent designer? If ordinary science fails to account for something as extraordinary as the sensation of light on the eye, how scientific is it to shut one's eyes to alternative views and deeper truths, and the chance that not everything shakes out right just because? "The contemporary argument for intelligent design is based on physical evidence and a straightforward application of logic," Behe insists. "In the absence of any convincing nondesign explanation, we are justified in thinking that real intelligent design was involved in life."

ID promoters are careful not to say who or what their posited designer may be, nor whether it's a he or a she or a S/He or an anonymous corporation in Delaware. "Intelligent design itself says nothing about the religious concept of a creator," Behe writes. For many scientists, the disavowal rings disingenuous. Behe's appeal is not really for greater fairness and open-mindedness, or a request that scientists delve more deeply and rigorously into the molecular basis of life than they have to date, conceiving more imaginative experiments, redoubling their efforts to find the perfect controls. The basic message of the designer school is, Sorry, folks, there's nothing more to be done. In the biology of molecules and cells, we've reached the limits of what science can tell us. We've reached a point of irreducible complexity, and if you can't reduce a complex object to simpler and more manageable parts, well, then, you can't do much with it, can you. Science requires some degree of reductionism, some picking apart and focusing on one or two variables at a time. But if natural selection supposedly couldn't concoct a clotting cascade piece by piece, what hope is there for science to trace it methodically back to the start?

Not only are molecular scientists unwilling to throw up their hands at any problem, and say, "Oy, it's too complicated! I've never seen anything so irreducibly complicated! How about if we just toss our lab notebooks in the autoclave, invoke the 'supernatural intervention' clause, and duck out for some fajitas and beer?" Scientists, as a group, are far too competitive and hardworking to say they can't do more when there obviously is so much more to be done. They also argue that the specific molecular assemblages and protein cascades cited by intelligent design advocates as being irreducibly complex and resistant to a Darwinian analysis can, with only moderate exertion, be disassembled into manageable subunits and those components explained as the products of natural selection. In *Finding Darwin's God: A Scientist's Search for Common Ground Between God and Evolution,* Kenneth Miller, a professor of biology at Brown University in Rhode Island, deconstructs many of ID's oft proffered instances of irreducible complexity. Among the most vivid is his vivisection of the choreography of blood clotting. He describes the step-by-step reactions that culminate in a clot: how trauma to the body's surface stimulates a succession of enzymes, or factors, circulating in the blood, each of them designated by a Roman numeral — for example, factor VIII, factor IX, factor X; and how the activation of one factor is contingent on the arousal of all the Roman soldiers preceding it; and how, at each node of the cascade, the strength of the biochemical signal is ramped up a millionfold; and how, finally, factor X bugles out a riotous reveille to an enzyme called thrombin, which clips little protective side chains off a ropy protein called fibrinogen, making the protein sticky. The newly gluey fibrinogens quickly ball together, and you've got your clot.

Miller admits that the scheme is intricate, a "Rube Goldberg machine," and that "if we take away any part of this system, we're in trouble." Medical geneticists have identified diseases stemming from mutations in just about every one of the factors in the clotting pathway, and they are all severe disorders. "No doubt about it. Clotting is an essential function, and it's not something to be messed with," Miller writes. "But does this also mean that it could not have evolved? Not at all."

As far as we know, Miller explains, the only animals that rely on a network of protein reactions to clot blood are vertebrates — we backboned mammals, birds, reptiles, amphibians, and fish — and some arthropods, particularly big, hard-shelled species like lobsters and crabs. But that doesn't mean a worm or a starfish will simply bleed to death if a blood vessel is severed. Creatures without clotting proteins rely instead on "sticky" white cells circulating in their bloodstream to patch them up. In the event of an injury, the sticky cells will cling to any proteins, like collagens, that jut out from the surface of the exposed skin; over a few minutes' time, enough white cells will have accreted at the gash site to form a plug that blocks further blood loss. Compared to the speed and elegance with which our clotting proteins operate, the sticky-cell Band-Aid approach is crude and slow. It can work only in creatures with relatively low blood pressure, which is what most invertebrates happen to have. Nevertheless, Miller argues, "It's exactly the kind of 'imperfect and simple' system that Darwin regarded as a starting point for evolution."

Lest you think even the invertebrate's simple system is too intricate to ascribe to evolutionary forces, Miller again demurs. Those white cells serve a variety of purposes other than clotting, including nutrient delivery. Imagine a blood vessel springing a leak, Miller suggests, and imagine that a few of these white cells have randomly acquired a mutation that make them sticky when exposed to the ragged, fibrous matrix of ruptured dermis. "Any change . . . in the white cells that made them stick, even just a little bit, to that foreign matrix of tissue proteins," he writes, "would be favored by natural selection because it would help to seal leaks." In other words, a random mutation that happened to lend the white cells of some ancestral worm or sea urchin a touch of Velcro would help turn current bleeder into future breeder, and so the mutation would be selected for and spread through the population, and, 'sblood! The rudiments of a clotting system are born.

Our vertebrate clotting mechanism relies on blood proteins rather than on whole cells to make clots, but still, the same sanguine logic applies. The protein factors that thicken our blood are very similar to proteins found in the pancreas and other organs that have nothing to do with clotting but instead clip and splice a variety of biochemical signals. Clipping and splicing, though, is precisely the sort of seamstressing skill needed to cross-link blood at a crisis point and bar its hasty departure. By all appearances, our clotting proteins were recruited from preexisting ranks of more generalized processing enzymes, and the genes encoding the processing proteins duplicated to enlarge the talent pool. Gradually, a number of these processors, these so-called serine proteases, were committed to the task of clotting, their reflexes honed, their internal signaling network tightened and amplified and made mutually, obligately symbiotic, the fate and force of each bound up with the rest. Today, clotting is like professional baseball. Just as the Yankees can't play with an eight-person team, so the loss of just one clotting factor can threaten your life, knock you out of the game. The current interdependency of our clotting network accounts for its extraordinary speed and vigor, but that doesn't mean it was ever thus, or can be only thus. "Blood clotting is not an all-or-none phenomenon," writes Miller. "Like any complex system, it can begin to evolve, imperfect and simple, from the basic materials of blood and tissue." A sea urchin makes do with its simple white cells, and two kids with a ball can play catch in the park.

We don't know how life began. We don't know if it was physically inevitable, given Earth's geochemistry and the sun's generosity, and we certainly don't know if it was in any way spiritually inspired — an expression of divine love, or of cosmic curiosity, the universe's desire to understand itself. We don't know what the first life forms looked like or how they behaved. They might have been made of ribonucleic acid, RNA, or of proteins, or of molecules as yet undiscovered and unaccused. We don't know exactly when, after the formation of Earth 4.5 billion years ago, life first arose. It might well have been quite early in our planet's history. Harold Urey and Stanley Miller of the University of Chicago won international fame in the 1950s when they sought to recapitulate in the laboratory the conditions of early Earth and managed to generate amino acids, the building blocks of proteins. Miller was once asked to speculate on how long it might have taken life to originate. "A decade is probably too short, and so is a century," he replied. "But ten or a hundred thousand years seems OK, and if you can't do it in a million years, you probably can't do it at all." The operative verb in the above passage, however, is "speculate." The fossil evidence for early life is woefully, chasmically gapped. Whatever the biochemical nature of the matriarchal molecules that first managed to replicate themselves, they certainly had no hard parts, nothing for the sedimentary archives. Even after the self-copying chemicals succeeded in sealing themselves off from their surroundings, each one mapping out the boundary between me and not-me with a springy, lipidic membrane slicker and declaring itself a cell, still the young life gave no thought to tomorrow.

However life got started, one thing is clear. Life so loved being alive that it has never, since its sputtering start, for a moment ceased to live. Through the billions of years since the first cells arose, chubby bubbles enclosing the code for budding off more bubbles, life has carried on. The cipher of life, the text written in the nucleic phrases of DNA and RNA, is a universal code. Every living creature owns a piece of it. Every parasitic, periliving, propagandizing virus owns a piece of it. There is no other way of saying I'm alive but through the phonemes of nucleic acids. Had life arisen more than once, had its origins been polyphyletic rather than monophyletic, we'd see a multiplicity of codes, a selection of biochemical instructions for growth and maintenance. Yet we do not. We look at cells from creatures living on the ocean floor, 8,000 feet below the ocean surface, basking in the boiling plumes that seethe up through hydrothermal vents, and we see DNA. We pry open bacteria trapped in polar ice for more than a million years, and we see DNA. Species arise, multiply, diversify, and die, but DNA survives — if not in the spiny *Hallucigenia* of the Cambrian era, with its seven sets of clawed tentacles fit for scavenging the ocean floor, then in the predatory lungfish, *Dipterus*, of the Devonian age; if not in trilobites, then in pterodactyls; if not in dodos, then in Lewis Carroll. The timeline of life is segmented by major mass extinctions and minor mass extinctions, and in the worst of the die-offs, huge phyletic hanks were yanked off Earth, and the ranks of the vanquished outnumbered the hangers-on by a ratio of nine-plus to under one. No matter. DNA just kept repeating itself, over and under, somersaulting somewhere, in some cell, reading itself backward — AND never running dry.

Gunter Blobel, a cell biologist at Rockefeller University, Nobel laureate, and fair grist for a limerick, sees the plain splendor in life's unbroken tenure. "When it comes right down to it, you are not twenty or thirty or forty years old," he said. "You are 3.5 billion years old. Some people may say how terrible it is, this idea that we come from monkeys. Well, it's worse than that — or better, depending on your perspective. We come from cells from 3.5 billion years ago.

"There is this tremendous thread of life that goes back to when the first cells arose, and that will continue on after any of us die as individuals," he said. "It's continuous life, and continuous cell division, and we are all an extension of that continuity. Reincarnation and similar themes are poetic representations of biological reality."

If you want to see yourself as you really are, or as your ancestors were, or as your descendants will be, Blobel said, forget about the mirror. Crack open the cell and take a look inside.

7

Molecular Biology

Cells and Whistles

EVERY NIGHT BEFORE I go to bed, I grimly wage war in my mouth. First I floss, using three distinct products: normal slippery floss for most of the teeth, extrafine "floss on a stick" to get at the cramped back teeth, and the creepy concatenation of stiff and fluffy segments called "superfloss" for digging under the crowns and bridges. Then I deploy my plaque-removing, gum-massaging, erotically styled electronic toothbrush and brush for two minutes, longer if I decide to start folding laundry with my spare hand. Finally I rinse with a generous jigger of Listerine, swishing it cheek to cheek, round and round, the Bronx is up and the Battery's down, until all buccal and gingival decks have been swabbed with firewater, and I am free to spit.

Whenever I'm feeling glum or lazy and start thinking, Maybe tonight I'll skip a step or two, I rouse myself by recalling the vile day at age ten when my dentist told me that I had twenty-two new cavities and that he'd be spending every Saturday for the next six months with his distressingly furry forearms in my mouth; or my more recent realization, while gazing at my dental X-rays, that I'd had — Holy Novocain! can this be right? — nine root canals to date. Or I think of what I learned from Bonnie Bassler, the microbiologist at Princeton University, whose own thick pelt happens to reside on her head, about the floral story of tooth decay.

You probably know that cavities are caused by bacteria, she said. But what you may not appreciate is how sophisticated, resourceful, and relentlessly disciplined those bacteria can be. It turns out that, my tragic dental history notwithstanding, it isn't easy for the agents of decay to sink their teeth into ours, to remain in place long enough to drill holes through the tooth's protective enamel and thence to dine on the soft tissue beneath. For one thing, the mouth salivates constantly and purposefully: saliva is part of the body's defense system, a mildly antiseptic fluid designed to help sweep bacteria from the teeth and down into the gastric mulcher below. For another, tooth enamel is the hardest substance in the body. It is harder than bone, harder than an unclipped toenail on a camping trip. Enamel has enabled many a tooth to last posthumously into posterity; teeth are so abundantly represented in the fossil record, Michael Novacek of the American Museum of Natural History joked to me, that one might think the history of life on Earth consisted of teeth mating with teeth to beget other teeth.

So how do mouth bacteria manage to hang on and hammer through the enamel of live teeth, and in less than a single lifetime? We give them a head start by making poor dietary choices — chewing sugary bubblegum, for instance, or inexplicably reaching for one of those cellophane-wrapped hard candies that your grandmother has kept on her coffee table since the Ford administration. Not only does sugar attract bacteria, it helps them to cling to your teeth and begin their attack on your personal Pearly Harbor. The military analogy is a fitting one. Just as in a full-scale assault you bring out your bombers, your helicopters, your tanks, your Seabees and SEALs, so it is that six hundred distinct species of bacteria take part in the chop op. I'm not talking about six hundred individual bacteria. I'm talking about six hundred different species — or strains, as some microbiologists call them — each of them as genetically different from the other, said Bassler, "as Martians might be from humans." Hundreds of species, and hundreds of thousands or millions of members of those species, all cooperating to beat your teeth to a pulp. One species might be able to metabolize the sugar residues on the teeth, Bassler said, while another might be good at clinging to the enamel, and the next might release abrasive chemicals to begin scraping at the enamel. You can't see any of these wretched little grunts, of course. Like most bacterial cells, our oral flora are ridiculously small, a fraction the size of our body cells. Our pinhead, recall, could carry 3 million. But you can feel your cariel bacteria, oh yes, you can feel the thin coat of slime they leave on your teeth, the slime we call plaque. This plaque is Rasputin, or Mr. Johnson's cat. Do what you will, the plaque will always come back. "You can brush your teeth at night, but the bacteria will be back again by morning," said Bassler. "They'll be back, and not willy-nilly either. They will be back with the same highly structured order every time."

And so every day I hack right back, with floss and brush and minty rinse. I know the enemy. I admire the enemy. I may not keep the enemy away, but by systematically minimizing its impact on my teeth, I at least have a shot with the dentist.

You can't sterilize your mouth, or your hands or face, no matter how many bottles of Purell sanitizing gel you go through in a week. You are covered with bacteria. Maybe a half billion of them blanket your skin, a teeming microtropolis of several thousand different strains. Billions more happily fill the moist orifices of your body — the mouth of course, and nose, ears, vagina, urethra, anus, and lower intestines. When you breathe, you breathe in happenstance vortices of airborne bacteria, the great majority of which are harmless, are incapable of colonizing your lungs and making you sick. When you walk, you walk through and upon a drifting tulle of bacteria, like a Christo confabulation in Central Park but less saffrony. Rub your index finger across this page, and, poof, a million microbes ruffled or displaced. We galumph through all this life heedlessly, like giants in a Gary Larson cartoon, attending to it only when we seek to kill it — kill the plaque, the strep, the bearers of your tuba-toned bronchitis. Yet most bacteria are benign, want no more from us than we do from them, and many of them are quite useful, and some of them are essential to our survival. They feed us, they cook for us, they clean up our messes. By "fixing" nitrogen into a form fit for plants, root-dwelling bacteria help give plants leave to grow, and plants in turn give us all that we eat — our daily bread, our lettuce and tomato, our sliced roast beast. Once ingested, our meals are digested with the help of intestinal bacteria. Something like ninety-nine out of a hundred cells in our small intestine are bacterial cells, which flourish in the warmth and plenty of our plumbing and in return synthesize vitamins for us and help extract from our food essential nutrients that otherwise would pass through unclaimed.

Wherever you go, there they are, doughtily doing the world's dirty work. Dig up a gram of soil, a loamy pinch that can fit easily in a thimble, and you're looking at thousands of different species of bacteria, many of them detritus recyclers, breaking up the dumped and the dead and making them fit for new life. Or consider termites, the primary groundskeepers of tropical rainforests. They gnaw through dead or rotting trees and return much of the woody wealth back to the forest floor. What is a termite but a set of jaws joined to a petri dish, its gut a dense microecosystem of many hundreds of strains of microbes. Bacteria allow termites to wrest sustenance from sawdust and, like Geppetto, give dead wood a voice.

Some bacteria glitter, graced with the same incandescent chemicals that make a firefly glow; and just as a firefly flashes for love, so these luminiferous microbes will light up only when surrounded by others of their kind. Some bacteria play Jackson Pollock, spattering the stacked calcium outflows of Yellowstone National Park in streaks of pink, blue, green, amber, and brick — each color the signature of a distinct bacterial clan afeast on chalk soufflé.

Bacteria live everywhere, and in the most hellforsaken nowheres. They live on the summit of Mount Everest, and at the bottom of the sea; they live in polar icecaps and by boiling hydrothermal vents. They survive deep within rocks buried deep underground; they suck up heavy metals and oil spills and do laps in Love Canal. One bacterial species aptly named *Deinococcus radiodurans* can withstand a blast of radiation 1,500 times greater than the dose that would kill us, and 15 times greater than what would stir-fry that canonical survivor, the cockroach.

Yet as admirable as bacteria may be for their panplanetary powers and boundless vim, that brilliance ultimately redounds to a brilliance even grander, handier, and more foundational than theirs, the supreme brilliance of the entity of which bacteria and every other being on Earth is built: the cell. The cell is surely the greatest invention in the history of life on this planet, and ever since the first cell arose, as Gunter Blobel said, it has been all cell, all the time, a never-ending splitting of cells to make more cells, to keep life alive in the only way it knows how: in the context of the cell, by the bauplan of the cell. Bacteria exemplify the cellular nature of life because they are single-celled organisms. Each bacterium is a living being. It holds within itself the chemicals, components, and conditions required to sustain life, and it encapsulates the tremendous success story that is the cellular calling, the permanent nonpareil cellular vocation, which has never taken a vacation, never been out to lunch or out of line, or outgrown, outdone, outfoxed, outmoded, or rubbed out since the first cell arose some 3 billion years ago. This is the amazing thing, one of the most profound basic principles that biology offers: that once the first cell had pulled itself together, had assembled itself into a serviceable self-serving self, there was no turning back, and there has never been a cell-free moment since. Through ungodly long temporal caravans, stultifying passages of duffel bags stuffed with epochs, through ice ages and asteroid crashes, volcanic revolts, oceanic tantrums, and mass extinctions that destroyed 90 percent of life on Earth, still, not for a single day, a single nano- or pico- or atto-girl, yocto-boy second, has the world been without some scrim of cells somewhere, some thread of life, however threadbare, defiantly clinging to life. For the substrate stone, the stubborn cells may be so much plaque, the scum of the earth, and who knows but that the stone doesn't pine for a foolproof decellerator, a perfect Purell. Happily, it was not to be, in the history of our Earth; and aren't you glad the plaque always came back?

We know that all cells on Earth are monophyletic, are all descendants of a single founder cell, rather than being polyphyletic, of multiple, independent origins, because the unity of the genetic code tells us so. We also can see it in the structure of the cell, any cell, the cell of a bacterium, a corn plant, a fruit fly, a barfly. The cell, wherever it is stationed, has an unmistakable geography, a set of shared features that explain why it is the universal unit of life, and why it works so outrageously well. Think again about the Putumayo catalogue of bacteria we discussed: the ones in your mouth, the ones in your gut, the mountaineers, the thermophiles. In one sense, they're all very different from one another, every strain endowed with a subset of specialty genes that allows it to exploit weird resources like benzene or mercury and to weather the specific withering conditions of its niche. On the other hand, if you were to crack open any of these bacterial cells, you'd realize that they all look and feel very like-minded inside: similar chemical conditions, similar balance of acid and base. And the internal milieu of a bacterial cell is much like that of one of our liver or heart cells, or of any other cell of any other organism on Earth. This is the beauty and power of the cell, and one of the core insights to emerge from modern biology: A cell confronts the harshness and instability of the outside world by making itself a haven. A cell contains all the tools it needs to preserve order and stability within its borders, to keep its interior recesses warm and wet and chemically balanced. In this equilibrated, levelheaded setting, the cell's vast labor force of proteins and enzymes will operate at peak performance, and so sustain the cell in its state of mild grace. There is nothing more natural than a cell; the natural world, after all, is full of them. At the same time, a cell is the ultimate act of artifice, a climate-controlled limousine with cushioned seats and a private bar, cruising through a mad desert storm.

A cell is the basic unit of life, and the smallest unit of matter that can, by anyone's book, be considered alive. A virus is also a unit of matter that displays a few lifelike properties, most notably a zealous drive to replicate itself and the capacity to mutate and evolve; and a virus, being nothing more than a packet of genes wrapped in a jacket of protein and sugar molecules, is much smaller than even the smallest cells — bacterial cells. Nevertheless, most scientists argue that because a virus doesn't engage in such essential rituals of life as eating and excreting, and is entirely reliant on the apparatus of the host cell it infects to create new viral particles for it, a virus isn't a true life but a protolife, a wannabeing, a parasitic paralife as told on Post-it notes. They reserve their certificate of animate authenticity for the cell, as the smallest package of life on Earth and bearer of all the best gifts.

The cell lives, breathes, tastes, and makes waste, and when called upon will replicate. The cell is self-sufficient, and that is its conceptual beauty and power. But what, in a more practical, biomechanical, predilectional sense, is a cell? How does a cell work, what are its prime parts, and why is all life built on its spine? What does a cell look like, and why does it insist on being too small for the naked eye to see? First, I must point out that not every cell is microscopic. A cell has three basic parts to it: a greasy, waterproof outer membrane, the plasma membrane, which serves as the border between cell and setting, self and nonself; a gooey inner part, the cytoplasm, where most of the work of the cell is performed; and a cache of DNA, the cell's genetic content, its operating manual and ticket to tomorrow. In our cells, and the cells of any multicellular being and quite a few unicellular ones as well, the DNA is enclosed in a nucleus, a snug compartment surrounded by a smaller but double-layered version of the plasma membrane that moats the whole cell. In bacterial cells, the DNA floats free in the cytoplasm. Not surprisingly, the labels we assign to the two basic cell types lionize the DNA housing option we happen to possess. Cells with a nucleus are called eukaryotic cells, "eu" meaning "good" or "true," and "karyote" meaning "kernel" or "nucleus." Bacterial cells and other single-celled organisms lacking a nucleus we deride as "prokaryotic" — that's "pro" as in "pre," not "pro" as in "fan of" and certainly not as in "professional." Prokaryotic cells are "prenuclear" cells, the poor sods that kept the world busy for a billion years or so before the "good" cells, the ones with a nucleus, showed up. Bacteria have repaid the compliment through occasional displays of very professional pathogenicity — *viz.*, bacterial diseases such as bubonic plague, anthrax, syphilis, childbed fever, and, of course, tooth decay.

Nucleated or not, cells consist of these three defining ingredients, and it so happens that one type of bioentity that fulfills the criteria of cellhood is the egg. An egg has an outer membrane, a viscous cytoplasm that in the edible egg we call the yolk, and a set of genes — only half the number of genes needed to spawn an offspring, and half the number of genes found in other body cells of the egg bearer, but a gene set nonetheless. An egg, then, before it merges its DNA with a gene set supplied by a sperm and starts developing into an embryo, is a single cell, and that goes for the egg you can see well enough to scramble. Yes, believe it or not, an unfertilized chicken egg of the kind you buy at the grocery store is a single cell, although strictly speaking it's the cheery, marmalade-colored yolk of the egg that is bounded by the plasma membrane and thus qualifies as the cell proper. The translucent, whippable, protein-rich "egg white," the hard outer shell of calcium chloride, and the thin, slippery membrane lining the shell are all bonus coats added on later, as the yolk makes its way down the mother's cloaca. Still, chicken yolks are no joke, and they keep getting ever more jumbo even as we fret over the wisdom of eating any eggs at all. The largest egg in the world, and thus the largest cell in the world, is the ostrich egg, which measures about eight by five inches and weighs three pounds with its extracellular shell, two pounds without. (Interestingly, the ostrich egg is also the smallest bird egg relative to the size of its mother, amounting to only 1 percent of the female ostrich's body mass. The she-birds most deserving of every mother's pity are the kiwis and hummingbirds, which lay eggs that are 25 percent as big as they are — the equivalent of a woman giving birth to a thirty-pound baby.)

There are other cells in this world that can be sized up by the bald eyeball. Most bacterial strains are decidedly microbial, in the range of a millionth of a meter across, but *Thiomargarita namibiensis*, a sulfur-loving bacterium first discovered off the coast of Namibia, is a defiant millimeter wide, as big as the period you're about to reach. Among the so-called protozoa, a ragtag phylum of single-celled and usually invisible creatures which includes such laboratory staples as the amoeba and the paramecium, we can also find a handful of outlying colossi. Largest by far are the foraminifers, ocean-dwelling protozoa that may grow to a length of two inches; like bird eggs, these overachieving unicells are encased in hard outer shells, a mothering matrix of crystalline calcite that each foramin forms for itself.

Yet such macrobial cells are the exception, and the vast bulk of the world's biomass is built of motes. Our cells, an elephant's cells, the cells of the largest animal that has ever lived — a female blue whale — are tiny, an average of $1/2{,}500$ of an inch across. What's so great about being so small? I asked many of the biologists I interviewed. Why cells? Why build bodies, no matter how big they're destined to be, of parts too tiny to see? Why shouldn't we be made of what we seem to be made of — large sheets of unified matter, of tissue layers slathered one on top of another?

To be small is to have control, Cynthia Wolberger of Johns Hopkins University told me. Small is manageable. Small is flexible. The cell is shielded from its environment and so can control what happens inside in a way it cannot control the world outside. And the smaller the space in need of oversight, the tighter and sharper and more dynamic that control can be.

Companies have learned this lesson again and again, of the inherent verve and elasticity of the small, close-knit, semiautonomous team. So long as your individual fiefdoms remain compact and defined, your corporation can retain the cunning of David even as it assumes the multinational grasp of Goliath. We multicellular beings can obviously grow huge as well, all the while remaining biochemically nimble and shielded from the vagaries of our volatile world, because we are constructed of manageably modest parts.

The best way to understand the benefits of smallness to a cell, Wolberger said, is by taking a quick look inside. And here, it must be said, is where the picture gets a little bit ugly. I asked Wolberger what the cell would look like if it were blown up to the dimension of a desktop accessory.

Without a moment's hesitation, she replied gaily, "It would look like snot."

Snot?

"Yes, cells are very gooey and viscous," she said. "We do a lot of experiments in vitro, in a test tube, isolating elements of a cell in what is essentially a glass of water with salt and a chemical buffer. I like to remind my students that in vivo, in the real conditions of the cell, things are much thicker and more syrupy than that. They're more like snot."

On top of this unappetizing imagery, we have the offputting thickness of cellular nomenclature. You may be the proud possessor of 74 trillion cells, but the jargon of cell biology can make you feel like an alien without a green card or city map. Breach the border of the plasma membrane, and, whoops, you're smack up against the rough endoplasmic reticulum, a series of flattened sacs where proteins are made; or scraping along the Golgi apparatus, another stack of flattened sacs, where proteins are stored or chemically adjusted as needed; or whap, whap, whapping across the vesicles, the lysosomes, the ribosomes, the mitochondria. Even the umbrella term for the many little structures of the cell, "organelles," seems unnecessarily officious.

Never mind. Do not be deterred by either the in vivo viscosity or the verbal pomposity. The world of the cell is really not so different from our own. Cells may be small, roughly halfway between the size of an adult human and the size of an atom, but they behave more along the lines of classical, Newtonian, pushmi-pullyu everyday physics than by the foggy, probabilistic rules of quantum mechanics, where electrons vanish from one orbital lane and pop back up in another. Even the tiniest cells are shapely and fully 3-D, and though the basic cell morph may tend toward, shall we say, lava lamp droppings, specialized cells may assume specialized, elegant forms. Seen through a microscope, skin cells look like dinner plates fit for stacking, red blood cells like New York bialys, liver cells like shoeboxes lined up on their sides. Cells of the body normally stay put and obey the rules conveyed by the ambient chemical signals of the organ they're part of, but all cells at bottom are as fierce and twitchy as cats. Cut a few cells away from, say, a kidney, heart, or tongue, place them in a petri dish with a slick of broth and the right nutrients, and the cells will begin crawling like sovereign zootica, creatures of the Precambrian seabed. Watch the cells through a microscope and see how they thrust their edges out wide, like the wings of a bat or the fins of a manta ray, and how they drag themselves forward in search of more food, and cringe and rear back at the touch of another wandering cell. Cells are so strong that you wonder their owners can ever feel weak. Ants are famously heavy lifters, able to carry loads ten, twenty times their size; but cells, declares Scott Fraser, a bioengineer at Caltech, are at least one step up from the ants. In studies that use laser tweezers and plastic beads to explore how cells signal each other, a cell in a culture dish will grab at the beads by wrapping a bit of its plasma membrane around them, and then yank them free of the tweezers, an act not unlike a human uprooting a tree.

Cells are the unit of life, and they preen about it, throb with it, in every pore and ruffle. And the units of uppermost note in the unit of life, the molecules that do all the work of the cell, the moving, the shaking, the yanking, the eating and excreting, and the making of new movers and shakers of every make, are the proteins. Understanding the cell means understanding proteins, and this brings us to a minor point that many biologists admitted they find persistently frustrating: the public's narrow idea of what a protein is. Stephen Mayo is a professor at Caltech who runs a laboratory at the Broad Center for Biological Sciences, one of the newer buildings on campus and one of the few with an elaborate security system to prevent the theft of one or another $100,000 piece of equipment. He is young, tall, trim, dressed in crisp chinos and a striped tailored shirt rolled up at the sleeves. Mayo's office is spacious and sunny and understatedly luxurious, a reflection of the vast economic potential that his biomedical research is thought to hold. Mayo is trying to design new proteins that might in turn be incorporated into new drugs. Sometimes he attends or hosts social events with his wife, who is a volunteer with the Junior League, and he meets people in all sorts of professions. "When they ask me, 'What do you do for a living?' I take a deep breath," he said. "I tell them that I run a lab at a university and that we work on proteins. 'Oh,' they say. 'So you're a nutritionist?' People hear the word 'protein,' and the first thing they think of is hamburger." He'll explain to them that, no, he's trying to develop computer technology to design new proteins, new biological molecules for use in medical and pharmaceutical products. "But all the time I can see that in the back of their mind, they're still thinking about hamburgers," Mayo said. "They're wondering, What's wrong with the hamburger I'm about to be served?"

There is, of course, a connection between the protein in hamburgers, and the proteins to which the Mayo team is devoted. When you eat meat, you are eating cells, and cells are full of proteins. When you eat broccoli, you're also eating cells that are full of proteins. Our bodies need a steady supply of dietary protein to build new cells, repair damaged ones, replenish the immune system, and otherwise keep all the parts powered. The reason why hamburger is so much more readily linked to the categorical term "protein" than is steamed broccoli is that animal meat, which is made of muscle cells, is a denser source of protein and because those proteins more closely resemble our own. Hence, it's quicker and easier to obtain the protein components essential to our fleshly upkeep by devouring the flesh of another animal than by reaching for a peach, although as any vegetarian can attest, the plant kingdom is vast and varied and, with reasonable attention to dietary details, you can accrue all the protein you need from somewhere beneath its verdant canopy.

Whatever their source, dietary proteins are dull and lifeless things, and a sad, blinkered way to view the proteins of which Mayo and other biologists speak. What does the stomach do, after all, but tear any proteins encountered into the smallest possible bits, deactivating, desecrating, and *denaturing* them, as a protein chemist might put it. That is the stomach's job, to flatten a meal so it can be scavenged for spare parts. Let's chuck the steak as synecdoche for the molecule. Proteins are so much more than dead meat.

What then is a protein, in its natural state, on its rightful cellular stage? Technically, a protein is a string of amino acids, distinctive clusters made primarily of the elements most strongly associated with life — carbon, oxygen, hydrogen, and nitrogen — arrayed in a style that lends each amino acid a little knob of positive charge and a little knob of negative charge. That characteristic of molecular bipolarism, of carrying a duality of charges, makes amino acids ideal for linking together into a great diversity of structures, just as the holes and pegs of Lego pieces allow them to be snapped into model drawbridges, Ferris wheels, dinosaurs, and other marvels displayed on the cover of the Lego box, if not on your living room floor. Cells either synthesize amino acids from scratch or extract them from food, and then link the chemical subunits together to fashion a fresh protein supply. Those proteins differ considerably in size, from trinkets called peptides that are a couple of dozen amino acids long, to tumbling, operatic chains of several thousand amino acids. Keep in mind that "small" and "large" are relative terms here, and even the bulkiest proteins are still maybe a hundred-thousandth the size of a sesame seed.

Far more important than a protein's size is its shape, how its chain of amino acids folds, curls, puckers, and zags in three-dimensional space. Proteins are often described as being little "machines" in the cell, but that industrial, boxy term belies their Jean Arp curviness and Breck girl bounce. If you could watch proteins tumbling across your desk, they might look like an exceptionally stylish collection of Nerf balls, or origami animals made of butter and clay. And though proteins come in many textures, if you could touch a typical one, press down on it with an index finger, it would feel, in keeping with its cellular locale, viscous and mucousy on top, but with a decided firmness underneath. There is nothing silly about this protein putty, nothing slapdash about a protein's form, for from a protein's form its function follows. The specifics of a protein's shape, and the way its positive and negative electric charges are distributed along its contours, are what allow each protein to carry out its allotted tasks. A single cell might have 50,000 different proteins in its borders, some with deep notches, others with fingerlings thrust out in victorious V's, still others bearing confettilike streamers designed to wrap around a target molecule in a helical hug, or a combination of these and other recurring protein motifs. Most proteins have their stiff parts and their flexible parts, regions that remain relatively fixed throughout the life of the protein, and portions that respond to prodding from neighboring molecules and will shape-shift and switch tasks accordingly. Turn, turn, turn. Proteins live to work, and they live in a place much like Manhattan, a teeming city that never sleeps, where all that counts is how you look, and what you do.

What is it that proteins so busily, prodigiously do? Most of them are enzymes, proteins that help activate or accelerate chemical reactions in the cell by bringing together ingredients that might otherwise remain separate, or that change the shape of other proteins and hence prompt them to venture forth and ignite a chemical reaction. The distinctive structure of an enzyme is designed to fit smartly with only one or a handful of target molecules in the cell, the way your cell phone will snap only into its official recharging device but not the one of your parents, spouse, or anybody else in your immediate zip code. Once the enzyme has coupled with its target, its substrate, it can fulfill its specialized transformative mission. For example, there are enzymes in liver cells that are shaped to recognize rings of cholesterol, and once they have latched onto the greasy circlets, they help suture them into necessary sex hormones like testosterone and estrogen. Other liver enzymes conjoin salts, acids, cholesterol, fats, and pigments into the bitter yellow-brown digestive brew called bile. Then there is the face-saving liver enzyme, alcohol dehydrogenase, which helps break down the alcohol molecules in your blender drink into smaller, nonintoxicating pieces before you have a chance to pass out, throw up, or begin impersonating Peggy Lee.

And that's not all there is, my friend, that keeps us dancing. Enzymes in white blood cells can dissolve away viral shells, enzymes in our pancreatic cells help police how much sugar thickens our blood, enzymes in nerve cells make the chemical signals that stream through our brain and allow us to think, feel, do things, regret doing those things instead of other things, and fill prescriptions for Effexor.

Apart from the straightforward enzymes, there are the structural proteins that form the cell's filamentous supportive matrix called the cytoskeleton, "cyto" being the Latin word for "cell." Like bones, structural proteins give the cell its shape and integrity, and like bone tissue they are not at all inert, are in fact so feisty and eager to flaunt their powers that one might think they belonged to the metaphoric skeletons that one tries to keep in one's closet. Most renowned among the structural proteins is actin, found in all eukaryotic cells, a versatile molecule that not only serves as material for the cell's beams and girders, but also operates in a transportation capacity, shuttling other cell proteins from place to place, or helping to haul the trash out of the cell and dump it into the bloodstream, or putting everything into position during that most delicate and complex of cellular maneuvers, the splitting of one cell into two. In muscle cells, actin cooperates with another structural protein, myosin, to pull muscle cells in during a contractile motion, like the flexing of biceps or the squeezing of a bolus of food down the throat, and to relax the muscle fiber when the curl or swallow is through.

Because structural proteins can be as bustling and busybodied as any of the body's textbook pyrotechnic enzymes, some scientists argue that all proteins are enzymes, are engines of change, and life, and levity. The word "enzyme" means "to leaven," an etymological tip of the hat to the yeasty proteins that leaven bread, and wine, and thou. "L'chaim," too, is thought to share etymological roots with "zyme." It is the Hebrew toast to be spoken at a feast, over an alcoholic beverage of the toaster's choosing, and it means, quite simply, "To life."

The cell, then, bristles with proteins, with enzymes, with life. If you could peel the top off the cell for an insider's view, said Tom Maniatis, a biologist at Harvard University, it would be like looking into an ant pile or beehive, but at fast-forward speed. "There would be frantic activity, with things moving in every direction, and molecules being transported from place to place at a lightning pace." Picture as well a lot of darting through doorways and giant silent suckings, and vanishings into thick air. The membranes that girdle the cell and the nucleus are pocked with pores and channels that open and shut, and molecules enter and molecules escape; and everywhere in the cytoplasm there are bubblelike baggies called vesicles that bobble along the cell's railway tracks of actin and approach molecules and tackle them with the straitjacket of themselves, and take them to a new locale and then spit them back out; and there are other, ghastlier sacks, the lysosomes, the little stomachs of the cell, filled with blistering acids that destroy whatever cytotrash they suck inside. And in this panting antsy atelier, this hive in hyperdrive, many protein players travel as gangs, as bulbous protein complexes of three, six, a dozen diverse enzyme talents locked together through their structural complementarity, the happy meshing of knobs and clefts, positives and negatives. Until very recently, biologists tended to think of proteins in isolation, as rugged individualists, a set of singletons or monomers in the cell single-mindedly doing their jobs. One of the major insights of the last few years, whose magnitude continues to mount the more we study the in vivo vales of the cell, is that most proteins operate in teams, as polyproteins, and that the result of their pooled talents may be radically different from what one might have predicted by considering their enzymatic properties independently. What's more, the protein-protein allegiances are fluid and fungible. A protein may insinuate itself into one polypod one moment, and then break away and join forces with another protein set the next, and then another a few seconds or minutes or days later — fulfilling a different enzymatic mandate with each collegial union. And nowhere is protein clubbiness more evident than in matters of the family business: the procreation of proteins and of the busy, indefatigable, sometimes divisible republic on which they stand.

Tom Maniatis said that if you could strip the top off a eukaryotic cell, you'd see a tumult of activity; it's time to walk through the nuclear door, to the most rambunctious rumpus hall of all.

By popular reputation, DNA is a colossus among molecules, and certainly the bearer of many aliases. Our DNA is our genes, the things we inherit half from Mother, half from Dad, and that we offhandedly blame for our bad teeth, or our inability to separate light laundry from dark. Our DNA is our chromosomes, or our baby's chromosomes, the twenty-three pairs of little sausage-shaped bodies, as crimped and bent and acrobatic as Keith Haring cartoons, that are isolated in a prenatal amniocentesis test, stained, and finally scanned for signs of troubling breaks, imperfections, or duplications. Human DNA is also known as the human genome, star of the eponymous Human Genome Project, the multinational, multibillion-dollar effort to "map and sequence" the entire human genetic code, to specify every one of the 3 billion chemical letters of which human DNA is composed. DNA has been exalted to an almost idolatrous degree, the Holy Grail turned golden calf, and it now has the opposite problem of protein. Whereas a protein is seen as just another ingredient in meat, DNA feels far too storied for everyday bodies, and too big or too dangerous to eat. How else to explain the common misconception that the only grocery products with any genes in them are "genetically modified" foods, and the effort of some restaurateurs to advertise their refusal to use genetically modified foods in their kitchens by displaying on their menus a little picture of a double helix with a red slash drawn across it?

Of course, while you likely never give it a thought, you eat DNA all the time, even if your food is certified organic, the product solely of traditional means of genetic modification — i.e., the selective plant breeding, cross-hybridization, and animal husbandry techniques that humans have employed over the last 10,000 years — and guaranteed free of the contemporary "Frankenfood" finaglings that might have installed specific genes to confer on the crop resistance to, say, frost or fungus. Even then, while eating organic food, you swallow billions of genes arrayed along millions of DNA molecules every day. Every time you eat steak, you're eating a slab of cow muscle, composed of millions of cow cells, and those cells are full of protein, myosin and actin aplenty and more; and the plasma membrane around each cell and the smaller nuclear membrane within each cell are bubbles of cholesterol, and in the middle of each cow nucleus is cow DNA, the complete set of cow genes distributed on the thirty pairs of cow chromosomes that constitute the cow genome. From each cow genome enthroned in each cow cell, an entirely new cow, a Dolly-like clone of the donor cow, can be grown, giving fresh meaning to the claim that you are so hungry you could "eat a cow." You eat potato DNA, bean genomes, tomato chromosomes; if your diet is healthfully varied, you've devoured the source codes for thousands of species in your lifetime. Genomes are not just the stuff of high-concept, high-priced projects. They are stuffed into all our body cells, unabridged copies of the DNA molecule that each of our parents had bequeathed to us in demipart at the moment of our conception, and that the briskly proliferating cells during fetal development had replicated and bequeathed to every daughter cell, and that our adult cells still hold and still solemnly copy every time they divide. The only cells without DNA are our circulating red blood cells, the specialized cells that carry oxygen throughout the body. Red blood cells develop in our bone marrow from precursor cells that do have DNA in them, but in the last stage of their maturation, when they are ready to serve as our breath of life, as the emissaries from the lungs to every far-flung cell we own, the red blood cells spit out their nucleus and its DNA molecule to leave plenty of room for the activities of their resident hemoglobin proteins, the molecules that capture oxygen.

This is perhaps the premier point in the story of DNA: that in almost every cell of our body can be found a personal copy of the complete DNA molecule, with all our genetic information, all twenty-three pairs of our chromosomes, all our genes, and all the lengthy filler between genes, all 3 billion bits that constitute our very own human genome. It may not be *the* Human Genome, the one that scientists have largely finished mapping and sequencing; that official map is based on a compendium of genetic samples extracted from a handful of people, including patients in long-standing, important genetics studies and a couple of scientists with a long-standing sense of self-importance. Our own human genomes, though, the humble ones tucked into the nuclei of nearly all of our cells, are very similar to the great Human Genome spelled out in the databanks of the National Institutes of Health and other research centers. We human beings are, genetically, 99.9 percent identical one to another. Those few places where our genomes differ — from the archived archetype and from one another — help explain the individual differences that our eyes easily seize on, and too easily magnify. If only we could see the genomes we carry within us; then we might appreciate the homogeneous depths of our common humanity.

Still, nothing can be as familiar to us as the genome we carry, for it is photocopied *in toto* into every nucleated cell of the body. Our liver cells may make enzymes to detoxify alcohol, and our white blood cells specialize in decapitating microbes, but at their core is the same DNA molecule, same genome, same chromosomes, same set of genes. Where the DNA of a liver cell differs from that found in kidney or bone is in how the molecule is mollycoddled by the protein company that it keeps.

To understand the dynamics between DNA clonality and protein heterodoxy, we must look closer at the pampered colossus on its nuclear divan. DNA is a molecule that, if stretched out, would be as tall as a kindergarten child; but even in its supercompressed state inside the microscopic nucleus, still DNA is hundreds of times the size of an average protein. Yet for all its bulk, DNA is ultimately a simple molecule, far simpler, in fact, than many of the proteins that surround it. Whereas proteins are constructed from twenty different types of subunits, twenty different amino acids from which to pick, mix, and patch, DNA, amazingly enough, makes do with just four different chemical modules, called bases, underlying its frame. Formally, the four are cytosine, guanine, adenine, and thymine, but, as with presidents and fashion designers, their initials — C, G, A, and T — usually suffice. Each of the four bases is a distinctive but also relatively simple construct of nitrogen and carbon rings, which are tacked onto a spiraling backbone of sugar and phosphate molecules. The nitrogen and carbon ringlings project outward from their backbone and search for companionship. DNA is a double helix, after all, which means it consists of two strands of bases attached to two sugar-phosphate spinal columns. The Cs, Gs, As, and Ts on one strand face their counterpart bases on the other strand, and are held face to face by the gentle tugging of a hydrogen bond. But the pairing of bases across the strands is not arbitrary: A is always partnered with T, and G with C. This is the complementary match that feels right, that allows the DNA molecule to settle down and maintain structural integrity and uniformity up and down its span. Adenines and guanines are both relatively large bases, while thymine and cytosine are relatively small. You match husky partners with petite partners, and you get a nice clean vertical lineup. Isn't that sweet? Big like a male, little like a female. You might as well usher these complementary couples, these base pairs, right up the plank of Noah's ark.

DNA, then, is a two-faced molecule, two corkscrewing chemical chains loosely but lovingly locked in comfortable complementarity. On one side, a strand of 3 billion bases, millions upon millions of Cs, Gs, As, and Ts, arrayed in varying patterns, with a plethora of CATs and

TAGs and ACTs and TATAs! and long, long stuttering stretches of Ts or As or GCs repeating themselves until you're ready to GAG. And on the facing strand, the complementary lineup of 3 billion bases, so that where there's a CAT on one strand you'll find GTA on the other. The aim of the Human Genome Project was to determine the precise chemical sequence of all 3 billion base pairs of human DNA, and let me tell you it was a grueling and often tedious task, for much of the genome proved to be dauntingly repetitive, a drab, seemingly pointless wasteland within us. Specifically, large regions of the human genome turn out to consist of what is often referred to as "junk DNA," filler bases that seem to play little role in the molecule's primary assignment, of encoding the rules for making new proteins and new strands of DNA. We still don't know whether the apparent junk is truly junk, and persists in the bosom of our DNA because it does no harm and therefore the cell is under no pressure to eject it, or whether the junk plays still hidden but nonetheless essential roles, for example, by helping DNA to bend in all the right places, or as fodder for future evolutionary gambits. What we do know is that only a tiny fraction of the 3 billion base pairs, something like 5 to 10 percent, is devoted to the pressing biobusiness of protein production. Only 10 percent of our DNA, in other words, is the stuff that we designate as genes.

Here, then, is what it comes down to, the genomic mystique, the "genes" that we say it's all in the: maybe 300 million bases, scattered among a cast of 3 billion. These key chemical sequences encode our body's proteins; they are the recipes, the formulas, the runes for those proteins. At its simplest, this is what a gene is: a recipe for a protein, written in the script of DNA, in a lineup of As, Cs, Ts, and Gs. The code works in triplets: three bases signify one amino acid. If the stretch of DNA says CAT, you are looking at the code for the amino acid histidine, which is felicitously abbreviated as the feline "his." If you see GTT, then you know, aha, the amino acid valine wanted here.* And there are punctuation marks, too, triplets that mean "protein recipe starts here" and triplets that, like a square bullet or a ###, tell you you've reached the recipe's end. Other codes are like the dynamics notations in a musical

* Because the set of possible three-part combinations of DNA (64) exceeds that of amino acids in need of encoding (20 standard ones), most amino acids are specified by several different triplets of As, Ts, Gs, and Cs. The amino acids arginine, leucine, and serine all claim the uppermost number of six descriptors apiece, while poor tryptophan and methionine are accorded just one monogram each. Not surprisingly, tryptophan ends up being a relatively rare subunit in the protein community, though nevertheless essential to human health and happiness. From tryptophan the body makes serotonin, the familiar brain chemical that drugs like Prozac seek to enhance.

score, saying, juice it up here, make lots of that protein, or soft-pedal there, just a couple will do.

But these protein playbooks are not at all straightforward or linearly organized. Different parts of a gene, different steps in the recipe, may be inscribed in very different parts of the DNA macromolecule, and only get "read" as a coherent narrative at the moment the protein is fashioned. Junk and nonsense abound, not just between genes, but within the genes, too.* Scientists may have largely completed the spelling out of the human genome, but the sequence is a bare-bones beginning, the opening ACT, and there's so much left to be divined from this gnostic epic poem. We're not even sure how many genes there are in human DNA; every time we scan the code more closely, the total drops. As recently as the late 1990s, the accepted figure for the approximate number of human genes was 100,000. By the turn of the millennium, it had fallen to 80,000. A couple of years later that number was slashed in half. The latest tally has us in the low- to mid-20,000s.

Yet the body has considerably more than 25,000 different proteins to its credit; by some estimates, there may be 200,000 at work in our cells. Obviously the dandy old rule of thumb that one gene equals one protein no longer stands: instead, genes are like the sentences "Eats shoots turkeys and ducks" or "Is there in truth no beauty?" or "Hereallyearnedforher." Change the spacing or punctuation of the sentence, and you upend its sense. Similarly, the body's genes obviously can be read many ways by the cell's sharp-eyed foremen — the proteins that sense the cell's need for new proteins and have the structural flair to latch onto the DNA molecule and jump-start the protein-making machine.

Let's say you are a pancreatic cell, and you have the misfortune of being hitched to an irrational organism who is angling for her ninth root canal by dipping repeatedly into her grandmother's candy dish. She unwraps and swallows three caramels in less than a minute. Her blood spikes with glucose. She needs a fresh batch of insulin — a protein that acts as a signal between cells and so is called a hormone — to stimulate her liver and muscle cells to mop up some of that excess blood sugar. The pancreas is the body's designated source of insulin, you are a working member of the pancreatic community, and you can't suddenly phone in diabetic; chances are you'll be expected to produce insulin.

* This is not true of the DNA in bacteria and other prokaryotes, which divide early and often and thus cannot afford to carry around the genetic equivalent of likes and ums and twiddling thumbs. Bacterial genomes are much cleaner and terser than our own.

How in the world do you do it? Luckily, you are a cell, the beneficiary of more than 3 billion years of evolutionary experience, and you know inherently what we *Homo sapiens* do not yet sapiens explicitly: every step needed to transmit a signal faithfully from the outside world, the extracellular environment, to the deepest tabernacle within, and to turn the signal into new protein. In rough sketch, though, this is what happens.

A pancreatic cell senses a need for its services when sugar molecules in the blood begin agitating and dimpling the cell's membrane. The distress signal is relayed down through the cytoplasm by squadrons of quickstepping, shape-shifting proteins. It's like a heartwarming Hollywood movie, in which a child writes a pleading letter to the president of the United States of America, and we see the envelope passing from the local postal clerk through the central post office along to the minions in the White House secretarial pool up to the assistants to the assistants and over to the president's outer ring of advisers, picking up a sense of flurried excitement at each stage, until finally, Should we show this letter to the president? Oh, yes, absolutely, the president must see this at once! The advisers burst into the suitably cellularly shaped Oval Office to find the president, as ever, surrounded on all sides — by bodyguards, legislators, lobbyists, dignitaries, indignitaries, the presidential physician, the presidential astrologer/personal trainer/hairstylist, and Ed Tatum from Omaha, Nebraska, who wandered in while looking for the bathroom. No matter the crowds. The letter bearers don't need the chief's undivided attention; like everyone else, they just want one little piece — a piece that they will, before conquering, quite expertly untangle.

DNA, as it sits in the nucleus, is a dense, matted hairball of nucleic acid strands cloaked in proteins and coiled and coiled and supercoiled again. Only when one cell is on the verge of splitting into two — as cells do with some frequency in high-turnover tissues like skin and blood, but only rarely in fairly stable organs like the brain — does the DNA then separate into the discernible bodies we call chromosomes. Otherwise, all the chromosomal pieces of the genome are fused and bunched together. In addition to being supercoiled, the DNA in an ordinary, nondividing cell, like our pancreatic cell with its epistolary subplot, is, of course, a double helix. It is a corkscrew configuration with one strand of bases matched up to a strand of complementary bases, and very chemically stable as a result. The relative toughness of the molecule explains why on rare occasions we can fish out samples of DNA from ancient sources, such as insects trapped in amber, and why the premise of *Jurassic Park* — that dinosaurs might be regrown from remnants of dinosaur genes found in fossils — is not terribly far-fetched.

Stability and utility, however, are two different matters. Just as a book must be opened to be read, so the relevant region of the double helix must be cleaved apart for its instructions to be understood. In that pancreatic cell, then, there are proteins that know where on the massive, twisted body of their resident DNA can be found the step-by-step recipe for piecing together more insulin. We can't yet say quite how the proteins know where to look amid the nuclear haystack of 3 billion base pairs, but we know they know, we know there are recognition proteins that are needle-nosed for the insulin code, because the pancreas, as a rule, makes insulin every day. On attaching to the proper position on the DNA, the proteins, the teams of proteins, gently unwind that region of the genome and separate the two strands of the helix, revealing two rows of bases bared like teeth in an open jaw. Now other proteins can glean from the exposed code the knowledge needed to create new insulin protein, and so give those archival teeth a voice. Of course, you don't want to start puttering around on the prized original document any more than you'd want to lend the original Declaration of Independence to a fifth-grade student for show-and-tell, or to a member of the House Ethics Committee for any reason whatsoever. Working too long and too vigorously with the exposed, dehelixed DNA risks introducing a mutation into the molecule, a structural defect that could lead to a few problems later on, like, say, cancer. As the first order of business, then, transcription proteins make a working chemical copy of the insulin gene, an RNA message of the gene — or messenger RNA, as scientists call it. They flutter along one fiber of the gaping DNA, and they read it tactilely, as a blind person reads Braille. They gather spare bases from elsewhere in the cell, and they piece together the RNA message, which looks a lot like the original gene, with one small exception: wherever the DNA code holds a thymine base, the transcription crew will install into their message a very close chemical cousin of thymine, the base uracil. Nice job! A fine first draft! Before the message is worthy of publication as protein, though, it must be revised by editing proteins, which deftly delete all the bits of filler code in the transcript and splice together the serious passages into a working formula for insulin.

That cleaned-up message is then delivered to one or more of the cell's many ribosomes, the spherical bundles of protein and RNA that synthesize all new protein merchandise. The ribosomes glide over the message and they, too, interpret it by touch. They scan the bases as triplets, every threesome the call letters for an amino acid, though here, in the argot of the cell's protein-making guild, it is not a CAT that yowls for histidine, but a CAU, not TGG that means the amino acid tryptophan, but UGG. The ribosomes read, scrounging for the requisite amino acids. Your cells are flea markets and yard sales filled with the building blocks for proteins and RNA and more proteins and new DNA. In the case of cobbling together insulin, the ribosomes need 110 amino acids. And when the pieces have all been stacked in a row, the artisans stand back, and let the new protein go. Sproing! Self-propelled, governed by an internal sense of proportion and purpose, the linear chain of amino acids folds and squirms and chases its tail and does the rumba and the ay, caramba, and, with very little help from the proteinous crowd around it, attains its three-dimensional Nerf ball origami form. This theatric transformation, from flat stack of amino acids to robust in-the-round protein, may happen with near spontaneity, driven by a thousand tiny pushes and pulls inherent in its constituent parts, but that doesn't mean it is child's play. Scientists remain baffled by the nuances of protein folding. They have become quite skillful at isolating and sequencing genes, and they have sequenced the entire genomes of many species beyond our own — of the mouse, fly, roundworm, rat, dog, horse, chimpanzee, a grim gallery of deadly pathogens. And with the DNA sequence of a gene in hand, they can say immediately what the amino acid chits of its protein "product" will be. Still, researchers can't predict from a genetic sequence or an amino acid sequence what the final, fully folded protein will look like, or what powers its contours will claim. That ignorance brings to mind Lewis Thomas's amusing meditation on how "deeply depressed" he'd be if he were told to do the job of his liver, and how he'd sooner take over the piloting of a 747 jet 40,000 feet over Denver. "Nothing would save me and my liver, if I were in charge," he wrote, "for I am, to face facts squarely, considerably less intelligent than my liver." Fortunately, the liver delivers without the good doctor's advice, and a newborn insulin protein needs neither understanding nor applause to find and fold on its own dotted lines and ready itself for duty in a wine-dark sea.

Every body is smarter than we are. For all the daunting complexity of protein synthesis, cells do it effortlessly, quickly, munificently. Often a single RNA message will be read by many ribosomes simultaneously, each reeling out its own copy of the protein. In the average human cell, some 2,000 new proteins are created every second, for a daily per-cell total of almost 173 million neonate proteins. Multiply that figure by the roughly 74 trillion cells in the human body, and you get a corpuswide quota of, egads, 1.28×10^{21} proteins manufactured each day. In light of this astonishing cellular productivity, why aren't we all just getting bigger and bigger? OK, we are, but this is no place to discuss the international obesity epidemic; besides, even the cells of hunter-gatherers whip up millions of trillions of new proteins a day, and look at how thin they are. The reason why our cells don't all swell and burst apart is that proceeding right in step with prodigious protein construction is ruthless protein destruction. Cells build proteins up, cells tear them back down again. A sizable number of a cell's proteins are enzymes devoted to the degradation of other proteins, including other degradative enzymes. There are enzymes that destroy collagen fibers, enzymes that destroy bone proteins, enzymes that destroy the enzymes that destroy collagen fibers and bone proteins. The average cellular protein survives only a day or two, and some perfectly good specimens emerge from their ribosomal birthing chamber and are instantly demolished.

All this protein turnover may seem terribly inefficient and wasteful. Why spend so much time eating the flesh and fiber of others only to have our cells spend so much time eating the flesh and fiber of themselves? Is the cell absurdly sloppy, absurdly perfectionist, or a contractor for the Pentagon? In truth, the constant protein churning illustrates a deep tenet of biology and brings us back to the question posed earlier of why cells are so small. Mary Kennedy, a neurobiologist at Caltech, explained it to me as the principle of "dynamic equilibrium," the idea that in a highly complex biological system like a cell, the pieces must fit together both precisely and loosely. An enzyme must fit the knobs and grooves of its intended target, but not the somewhat similar knobs and grooves of another molecule nearby. If the enzyme is supposed to attach to the section of the DNA molecule where the insulin gene is inscribed, for example, you don't want it attaching to the genetic sequence that holds the code for making thyroid hormone.

At the same time, you don't want the enzyme to stick to the DNA molecule at the insulin address and stay there as though bolted in place. You want the binding to be finicky but flexible, said Kennedy. Moreover, you want varying degrees of flexibility. Sometimes a protein will attach very firmly to its target, sometimes moderately so, sometimes barely so. And the relative commitment of the attachment itself conveys important information: I'm really holding on tight here, I'm serious about my assignment. I need a maximal output of insulin hormone. Or, I'm really just poking around here, window-shopping — there's no call for insulin output at the moment, but who knows what may happen tonight, after dessert. Maintaining a state of dynamic equilibrium, of looseness crossed with precision, said Kennedy, "allows you to have a huge amount of control and feedback at every level of the system." One way to sustain that specific squishiness is by crowding the cell's occupants but keeping them moving at the same time — having lots of proteins and RNA messages and the great cramped chromosomes all pressed shoulder to shoulder, but stirring and shifting position and in constant communication. It's rather like a subway car at rush hour. Passengers get in, passengers get off, some push their way to the middle of the car, others cluster around the doorways, people mutter, Excuse me, excuse me, as they elbow their way to the door and disembark before the ding-dong knells and the doors shut again. A couple of seats open up, and passengers standing nearby eye the opportunity and glance at each other to see who's in greatest need. Go ahead, you take it. No, no, please have a seat. I'm getting off in a couple of stops anyway, and besides, I'm younger and in much better shape. And though the system always seems on the verge of anarchy, I can tell you, as somebody who grew up riding the New York City subways, that in fact it's a miracle of frenzied efficiency, delivering millions of people to and from work each day over hundreds of miles of tracks, and rarely breaking down, and no matter how suffocatingly stuffed into a subway car I've been, I have always managed to wiggle my way doorward and have never missed my stop. The analogy is highly inexact, and I'm glad the subway isn't a cell, for a lot of the "passengers" getting off aren't headed to home or the office, but to the wrecking ball. This is the cell's way of maintaining lubricated, edgy motion: headily spooling out new RNA transcripts and proteins, steadily shredding the old.

Constant protein turnover also happens to be an excellent way of controlling protein behavior. Many proteins debut with a contingent expiration date stamped on their forehead: they are designed to fall apart rapidly unless a chemical signal from the outside intervenes and instructs them to do otherwise. This trick is particularly useful for keeping the cell's most powerful proteins in line, like those that prompt the cell to begin dividing. The idea here, said Susan Lindquist, a cell biologist and former director of the Whitehead Institute, is that you want growth-prone proteins on hand and ready to respond at a moment's notice, especially if you're an immune cell that might need to start replicating at the earliest viral provocation. At the same time, you don't want replication proteins loitering around the cell indefinitely, lest they start acting of their own accord and fomenting unwanted cell division. The solution: synthesize the proteins constantly, but make them unstable. Only when the appropriate growth hormones or other molecular envoys enter the cell and bind to the proteins are the proteins stabilized and put to work.

And so again we can see why a cell sets its sights beneath ours. The cell excels through micromanagement, and high protein density and perpetual protein flux are best accomplished aboard a tight, compact ship. With its waterproof membrane and modest dimensions, a cell can keep its proteins hemmed in, salts sorted out, pH optimized, dynamism equilibrated. Every cell is a stable community afeast on intrinsic instability, a living island like Manhattan, unto itself, but with all ears tuned to the world beyond.

The DNA inside your liver cells is identical to that inside the cells of your brain, tongue, pancreas, or bladder, and the DNA of every cell holds the instructions to do the work of any cell. Much of that work is routine and nonspecific, performed by every cell regardless of where it sits. All cells of the body must tap into their DNA codebook to make proteins that will crank the wheels of the Krebs cycle, for example, the stepwise transformation of food into usable cell fuel. All cells must also consult their DNA for making proteins to repair it whenever it gets broken or mutated; and sturdy though the molecule is, it needs upkeep every day.

But then there are the specialist codes, the protein formulas possessed by all cells, yet consulted by few. The genome within a bladder cell has the code for making insulin, and yet your bladder doesn't secrete insulin no matter how badly you need to pee. Your pancreas cells in theory can hammer out taste receptors to tell the bitter from the sweet, but the pancreas is a large, hammer-shaped gland hanging toward the rear of your abdominal cavity, and it has better things to do. Different cells of the body, then, differ somewhat in how their DNA behaves, in which genes are active, and which kept mum. Proteins are the laborers that do the fancy switchwork. They arouse genes, they stifle genes. Proteins in a brain cell will latch onto the DNA molecule and scan the code for making dopamine or serotonin, neurotransmitters that convey signals across our crinkled cortex. Why do brain cells have these proteins in their borders, while skin cells do not? How do brain cells know to make the proteins that home in on DNA and access the code for other brainy chemicals like serotonin and dopamine? If the DNA of your head cells is the same as the DNA of your toe cells, why can't you think with your feet, even when you try to vote with them?

Much of the answer to how cells differentiate and assume their tissue-specific identity remains hidden in the abiding mystery that is embryonic development. You started out as a single cell, a fertilized egg, and that potentate cell knew everything and saw to the farthest horizon and had the potential to give rise to all organs of the body. But as your embryo grew, its rapidly proliferating population of cells began budding off into discrete colonies, layers, sectors, primordial organs; and the greater the number of cells, the less freedom of motion and possibility each cell retained, the more committed to its location and vocation as a member of a limb or kidney or lung. In the course of differentiating, the genome within each cell underwent a series of subtle modifications. If the cell was destined to become part of the liver, genetic codes necessary for the production of bile and sex hormones would have been gently pressed into an active configuration, perhaps with the kinks of the DNA on which they sit being turned slightly outward, made accessible to the transcription proteins that give the codes voice. At the same time, genetic sequences of no use to a liver cell were muffled, tucked under and inward or slapped over with a few chemical "methyl groups," the cell's version of duct tape on the mouth. We understand little about embryonic development and the balletic genetics behind it, although extensive research on the subject is well under way, including on the much feted and politically freighted class of cells called stem cells — founder cells from which more specialized cell types then stem.

Yet even after our cells have assumed their basic identity, have been programmed in the art of acting like a smooth muscle cell or a hairy follicle cell, they continue to hone their skills and refresh their memories by listening to the voices around them. A liver cell knows it is a liver cell by dint of priming during embryogenesis, and because all the cells around it remind it of its liverishness every moment of every day. Cells are gossips, scolds, eavesdroppers, and sheep. They attend to their neighbors and hector their neighbors and keep one another in line. About half of the proteins in a cell are devoted to communications — to receiving signals from other cells and conveying advice and counsel back again. Cell membranes bristle with hundreds or thousands of receptor proteins, which protrude from the cell like outstretched arms or woven baskets or egg whisks. Each receptor type is shaped to embrace a particular molecule, a hormone, a growth factor, a song for the cell; and on meeting its designated match, the receptor protein shifts its shape in no uncertain terms to ensure that the entire viscous village beneath it hears. Cells send molecular missives across the tiny spaces of the extracellular matrix, and across the body buoyed in the blood or lymph. The cells of the pituitary gland at the base of the brain secrete sex hormones that persuade ovary cells to help ripen an egg, or testes cells to supply fresh sperm. The mast cells of the immune system, on encountering an invading allergen that they deem dangerous — mildew spore, cheap eye shadow dust, red alert! — will flood the surrounding tissue with the chemical histamine; and any cells in the vicinity blessed with histamine receptors will gamely react, yielding the swollen eyes, dripping sinuses, serial sneezing, and asthmatic wheezing that can make the body's inflammatory response so much nastier than the pathetic threat that spurred it.

Beyond chemical diplomacy, there is good old brute force. Cells are strong, as we've seen, stronger than ants, and they can tug and yank at the cells adjacent to them, or extrude from their surface their thin, long streamers, called philopodia, to deliver a few pointed pokes. That mechanical stimulation acts on the recipient cells much as a powerful hormone would, rearranging the cell's internal protein furniture and setting off a signaling cascade that tumbles right down to the nucleus. Through the medium of massage, a group of uncoordinated, introverted cells, each minding its own business at its preferred pace, in a flash can be rallied and synchronized and swayed to behave as one. When you cut yourself, it is the sensation of being yanked and stretched that spurs surrounding cells to begin dividing and heal the gap. Conversely, if a cell infected with a virus initiates its suicide program for the sake of the greater good, the urgent membrane ruffling that is a hallmark of programmed cell death can induce neighboring healthy cells to kill themselves, too, just in case.

Again and again scientists have seen cellular groupthink in action and eavesdropped on its propaganda machine. If you extract stem cells from an early mouse embryo and inject them into the bloodstream of an adult mouse, the fate of those ingénue cells will depend on where they land. Stem cells that lodge in the liver become liver cells, those trapped in a muscle get muscular, those caught in the kidney learn to go with the flow. Obviously the injected stem cells didn't have a chance to experience the hazing of normal embryonic development and whatever stepwise genetic changes it entails. Instead, each cell had to learn on the job, by osmosis, imitation, indoctrination. If the elder cells around it talked of nothing but a liver's lot — of secreting bile, regulating the blood supply, storing fats and sugars, detox duties — the stem cell absorbed the ambient information, imbibed hormones and other molecules designed to stimulate liver cells, and began responding as a liver cell would. Inside the nucleus of the stem cell, the DNA molecule would adapt itself to the specific demands of liver tissue. The pursuit of mastery never ends, and cell specialization requires lifelong education.

Cells must pay particular attention to their community when it comes to the monumental matter of division. Many cells of the body, as it happens, are primed to divide. Growth is their default position, the thing they will do unless told to do otherwise, and a lot of the chemical signals that cells send to one another are exactly that — signals of growth repression. Only on the removal of these inhibitory signals, coupled with the reception of positive signals that encourage growth, will a cell enter the tightly choreographed *ballet d'action* of division, a task carried out by a vast protein corps. The DNA molecule is pried apart, just as it is when its genes must be read, but this time the entire long-winded masterpiece is scanned, and a complementary copy made of all 3 billion exposed bases; and the copy is spell-checked for accuracy, and most of the mishaps repaired, at which point a matching strand can be made, and the newborn pair then entwined; and the two heavyweight molecules, mother DNA and her dutiful, duplicate daughter, are pulled over to opposite corners of the nucleus, and the nucleus is pinched down the middle into two little bubbles, each with its own copy of DNA, and the entire cell soon follows suit. Yes, just as cells love making protein, they love splitting up, and they do it quite well, in their well-controlled fashion. Millions of your body cells divide each day, and so your upper epidermis sloughs off and is replaced by new skin underneath, and your head hair grows half a foot a year, and your immune system can meet almost any pathogen it faces through the explosive expansion of suitable warrior cells.

Yet we live in the world, which may be the best of all possible worlds, but still it's not perfect. Every time a cell divides and its DNA is replicated, mistakes are made: a thymine is inserted where a guanine belongs, or a C base is inserted instead of the proper A; and really, what else would you expect in the course of copying a chemical text 3 billion nucleic letters long, which, if they were printed letters, would fill maybe 5,000 books the size of this one? Most of the mistakes in DNA replication are spotted by proofreading proteins and corrected before cell division is through; and of those few that slip through, most don't matter, for they fall into a harmless region of the genome. Once in a while, though, a serious mutation is overlooked and ends up in the final DNA script of the daughter cell, a change in the code that will yield a rotten, dysfunctional protein product somewhere down the line. And by far the rottenest proteins are those that "liberate" a cell from the constraints of its community, for they are the proteins that turn a cell cancerous. A cancer cell is a cell that is deaf to the chemical tutelage around it and indifferent to the slings and ruffles of its neighbors. It no longer needs hormonal inspiration from the outside to stabilize its stash of replication proteins but will make a set of the proteins and stabilize them of its own accord, and then make more and more and keep those, too. The receptors protruding from the surface of a cancer cell may be empty-handed above, but still their lower stalks shake and bend in the cytoplasm below and send shock waves through to the nucleus, with the order to grow, grow, grow. The sticky daubs on the outer membrane that keep healthy cells tethered together soften up until the cancer cell comes unglued, allowing it to travel where it pleases; and when the rebel cell settles down in new ground, still it hears nothing of the tissue around it, but listens only to its inner malignant hiss, telling it, You are a cell, and you must survive, and to survive, you must divide. But it is a false message, for in unhampered division, in its state of solipsistic genetic determinism, the cell kills the body and, with the body, itself.

The normal cells that we live with, the cells that abide by the laws and harmonics of multicellular existence, exemplify the dynamic equilibrium ever at work between a cell and its setting, or, zeroing in still further, between DNA and the proteins around it. Many biologists grumble about how DNA has been seriously misunderstood, stripped from its cellular context and petitioned for answers to everything — cancer, heart disease, bad moods, choice of mate. People talk about the nature-versus-nurture debate, and they want to know how much of who and what they are can be attributed to "nature," which is generally viewed as synonymous with their DNA, the specifics of their genetic code; and how much to "nurture" or "the environment," which commonly signifies the amorphous "world outside" and is characterized by variables like the childrearing practices and prejudices of their parents, or whether they attended an expensive, highly selective preschool or spent their formative years at the knee of Nanny Nickelodeon. Scientists have struggled mightily to impress on the public that the nature-nurture "debate" is dead, that it was an unscientific nonissue from the start, something pumped up and sustained by a media ever in love with conflict and horseraces. "It's unfortunate that there's a linguistic similarity between the words 'nature' and 'nurture,'" Stephen Jay Gould once lamented to me, for the euphonia alone "has helped keep this ill-formulated and misguided debate alive." You can't uncouple nature from nurture, he and other scientists insist, any more than you can uncouple a rectangle's length from its width. "It is a true union of influence," said Gould. "It's logically, mathematically, and philosophically impossible to pull them apart."

The result of all this assertive promotion of an interactionist rather than a dialectical perspective when it comes to dissecting the roots of human nature is the undoubtedly accurate but not terribly profound impression that, gee, maybe one's DNA and one's upbringing together help shape one's personality. In truth, the indivisible link between the two, between the instructions encoded in your genomic particulars, in your DNA, and the execution and interpretation of those instructions in real time, is profound, is embedded in the deep chemistry of every cell in your body. DNA may be called the master molecule, but it can do nothing on its own and must live through the proteins that subserve it; and those proteins attend closely and continuously to one another, and to the world around them, for clues to what they should make of their master. And on attending to external signals and looping back to the DNA, the proteins may sometimes change the very character of the genome, by subtly shifting which genes they activate, and how strongly, at any given time. Nature needs nurture, nurture kneads nature, and the codependent conversation never ends. It is ongoing everywhere within you. People often have the impression that if something is "encoded in their DNA," it must be static and unreachable. The environment, by contrast, is thought to be easily changed. Yet this impression is misleading. Your genome is not walled off from its setting. Every cell is a mad Manhattan microhabitat, and every genome a player in it. Genomes are responsive, open to change and modification.

Indeed, the pharmaceutical industry would love to tap into that genomic flexibility, to design drugs that would go right to the source code in a patient's cells, and fix a problem by fine-tuning gene expression — persuading the liver to make more of the high-density, or "good," form of cholesterol and less of the "bad" low-density lipoprotein, or bone tissue to remodel itself and heal a broken hip, or the brain to supply an optimized cocktail of neurochemicals to conquer depression, despair, a chronic sense of inadequacy, which may or may not be warranted but which is always unpleasant. And why not get rid of the recurrent nightmares, too, like the one in which you're onstage, dressed as the Scarecrow in *The Wizard of Oz,* and you can't remember the line that comes after "I would dance and be merry . . ."

Oh, for the day when we can have a heart-to-heart with our neurons and tailor our drugs to form-fit our genomes. Oh, to have the wits of a liver, the shrewdness of a single cell. Yes, my friend, life would be a . . . ding-a-derry! If we only had the brains.

8

Geology

Imagining World Pieces

WHEN YOU LIVE in the nation's capital, where every glorious monument to liberty is ringed with Jersey barriers and status is measured by the size not of one's paycheck or limousine but of one's Secret Service detail, you get used to imagining all sorts of disasters. A child's kite looks like a perfectly plausible anthrax distribution device. The van that screeches to a halt at an intersection when the light is still yellow surely houses a dirty bomb. A man dressed in a bulky, ill-fitting Brooks Brothers raincoat looks suspicious. A man *not* dressed in a bulky, ill-fitting Brooks Brothers raincoat looks suspicious.

Yes, in Washington, D.C., and its environs, you learn to think the unthinkable and to prepare for a wide range of emergencies, primarily by stockpiling duct tape, canned soup, and extra cat litter. One thing you almost never worry about, however, is an earthquake. Which is why, when I was working up in my office on a spring afternoon, and I felt the house start to jiggle and sway, I thought about everything other than an earthquake: a terrorist attack, a passing Abrams tank, my neighbor with the big dog and the Vin Diesel line of lawn tools, who runs his leaf blower at night, in the rain, just because, he's kindly explained to me, he can.

But as the shaking continued, I realized I'd felt the exact same sensation once before, when I lived in San Francisco, and that it could only be an earthquake. For nearly half a minute, the house shook, while I stayed stock-still in my chair, safely positioned beneath a large ceiling fan. I tried not to panic. I tried not to think about Carole King singing, "I Feel the Earth. Move. Under My Feet," but it was too late, and now maybe it is for you, too. Sorry! Finally, when I was sure at least the swaying part of the crisis had passed, I called my husband, whose office was in downtown Washington, some six miles away.

"Did you feel that?" I gasped.

"Feel what?"

"Well, you're not going to believe this," I said, "but I'm pretty sure we just had an earthquake."

"On some new medication, are we, dear?"

I muttered an imprecation, or maybe it was an oath, hung up the phone, and basked briefly in the warmth of righteous indignation. A moment later, my husband called back. "You're right," he said. He'd just seen a wire story go by, reporting an earthquake extending from parts of Virginia through to Maryland, where we live, and measuring 4.5 on the Richter scale. Venturing out into the hall, I saw that all the pictures on the wall were askew, and that one of them — a drawing of a woman who, come to think of it, looks a lot like Carole King — was on the verge of crashing to the floor.

The capital is not a seismically flamboyant region. It lacks California's tic-tac-toe of active fault lines, the lava luaus of Hawaii, the captious powder-point volcanoes found in that other place called Washington. Yet every now and again, even a sedate location with no known geological risk factors will give a sharp little shrug and demand that earth scientists pay it some mind. The intermittent jolts and tremors offer unambiguous evidence of a geological principle that truly merits the designation "bedrock": The planet we inhabit, the bedrock base on which we build our lives, is in a profound sense alive as well, animate from end to end and core to skin. Earth, as I said earlier, is often called the Goldilocks planet, where conditions are just right for life and it is neither too hot nor too cold, where atoms are free to form molecules and water droplets to pool into seas. There is something else about Goldilocks, beyond her exacting tastes, that makes her a noteworthy character, a fitting focus for our attentions. The girl cannot sit still. She's restless and impulsive and surprisingly rude. She wanders off into the woods without saying where she's headed or when she'll be home. She barges through doors uninvited, helps herself to everybody else's food, and breaks the furniture. But don't blame her. She can't help herself. Goldilocks is so raw and brilliant that she just has to let off some steam. Like Goldilocks the protagonist, Goldilocks the planet is a born dynamo, and without her constant twitching, humming, and seat bouncing, her intrinsic animation, Earth would not have any oceans, or skies, or buffers against the sun's full electromagnetic fury; and we animate beings, we DNA bearers, would never have picked ourselves up off the floor. The transaction was not one-sided, though. The restless, heave-hoing motions of the planet helped give rise to life, and restless life, in turn, reshaped Earth.

"We now understand that it isn't simply a matter of life accommodating to the buffets of physical change, but that life is a participant in the evolution of environments," said Andrew Knoll of Harvard University. "A grand theme of the history of our planet is how physical and biological Earth have coevolved through time."

When the whole world is your subject, it pays to be well-rounded, and geologists consider themselves the ultimate interdisciplinarians. They do fieldwork and lab work and crib from chemistry, physics, ecology, microbiology, botany, paleontology, complexity theory, mechanics, and, of course, computer modeling; geologists compete with protein chemists for their production of colorful computer-generated schematics that can be manipulated multifactorially in three-dimensional space and also make very nice screen savers. They love being outdoors, chipping away at rocks, leaping blithely from one sheer precipice to the next, and slowly acquiring the complexion of a Slim Jim. Geologists are often drawn to regions of great natural beauty and spotty safety records: active volcanoes, active fault lines, mountainous borders between sporadically warring nations. Unnatural eyesores can also have their appeal. When a new tunnel is blasted through a hillside, geologists will descend on the site for a chance to study the vast spans of geohistory that are fleetingly exposed, and if necessary may stall for time by throwing their graduate students in front of any oncoming cement mixers.

For geologists, every stone is a potential Rosetta stone, a key to a milestone moment in Earth's history, and to accompany a geologist through a park is to leave no stone unturned or outcrop unlearned. While strolling in the Arnold Arboretum on an unusually cold summer afternoon, Professor Kip Hodges, then at MIT, stopped at a thigh-high boulder that looked like a big lump of hardened cookie dough, and he took me through its résumé. "This is the kind of rock we commonly refer to as a conglomerate, which is just a rock that has blocks of different materials in it," Hodges said, pointing at the embedded chunks of what looked like nuts, or grayish white chocolate chips. "Notice that in this case you have blocks of many different sizes, and they're surrounded by a lot of fine-grained material, as though the blocks were just dumped and then locked into place." He ran his hand across the surface, and I followed suit. Very knobby, and chilly to the touch. When you see a mix of fine-grained and big material like this, Hodges explained, it's a good bet that the rock is of glacial origin. He took a seat on the boulder, and

I, less eagerly, followed suit. Very, *very* knobby, and most decidedly of glacial origin. The blocks that are now embedded in our exemplary boulder may have been rafted in by a creeping sheet of ice; the ice melted out, and the rocks became incorporated in the underlying sediment. "So the next question," Hodges said, "is when did all this happen?"

He then explained to me the considerable challenge of fixing an age to the boulder, a task that required, among other steps, sampling each embedded blockette for its relative concentration of radioactive tracers like uranium and thorium. But the effort proved exceptionally fruitful. The boulder on which we were perched turned out to be 570 to 590 million years old and just one of many like it discovered at sites around the world. Taken together with related research, the age and distribution of these rocks suggested that there had been a heretofore unknown ice age of great antiquity and scope, a hypothesis that many geologists are now pursuing. All of which demonstrates the geologist's first principle of fieldwork, Hodges said, as he gave the boulder an affectionate pat — that the real gems of the landscape are often the plainest-looking stones. It was a lesson I learned by the seat of my thin cotton pants.

"We live on a planet that records its own history," said Andrew Knoll. "I'm always amazed, when I drive across Utah, that I see this wonderful history unfolding before me, and you don't have to be a scientist to experience it. If you keep your eyes open as you go through the Grand Canyon, you'll see fossils. If you stop at almost any roadcut in the Midwest or look at the floors of almost any cathedral in Europe, you'll see fossils. It was hard to be a medieval penitent on one's knees without running into an ammonite every step of the way."

Yet for all the texts scratched onto its surface, Earth can also be a taciturn mule of a research subject, close to the vest and physically just about impenetrable. The deepest hole ever drilled got 7.6 miles down, a mere two-thousandths of the distance to the planet's searing inner core. Most of what geologists know about the inner earth they have gleaned indirectly. In the laboratory, they heat rocks and squeeze rocks and wring them into stone soup, charting the changes in the rocks' behavior and conductance properties with each new form of abuse. So armed, geologists can make the best of everyone else's bad day. When an earthquake strikes, they observe with the greatest possible precision how the waves of energy ripple outward from the quake's epicenter — their speed and direction, the relative decline of magnitude over distance, any harmonic overlays and reverberations they may have. The researchers can then compare the characteristics of those seismic waves with what they have learned about the conductance properties of different types of rock in their solid and molten states. Earthquakes, in other words, are like sonograms, the waveforms of seismic energy offering a portal into the imperial organs below.

Geologists sometimes complain that we have devoted more time and effort to exploring other planets than we have to our own, and they've been driven to propose extreme remedies for the knowledge gap. David Stevenson of Caltech, for example, has suggested that we make a slender crack right down to the earth's midpoint and then send in probes to sample the core directly, an idea he published in the scientific journal *Nature* under the Swiftian title "A Modest Proposal."

"I was being somewhat tongue-in-cheek, but I wanted people to realize that the idea may not be completely ridiculous," he said to me. "Making the crack in the first place could be difficult, but once you got it started, it would propagate under the effect of gravity."

Whatever technical limitations they may chafe under, geologists have come a long way since Jules Verne imagined the center of the earth as a kind of daffy reliquarium filled with mastodons, icthyosaurs, plesiosaurs, "Ape Gigans," and other members of nature's backlist. They've made their way around to something like the fifteenth-century fantasies of the good fire-breathing friar Savonarola.

We're all familiar with the idea that the earth's surface is broken into pieces, or tectonic plates, and that the movements of these plates has something to do with earthquakes, volcanic eruptions, and the pumice stone now gathering fungal spores in the corner of your shower stall. A simple glance at a desktop globe reveals that the plates have been slumming about for some time: South America and Africa look like matched pieces of a jigsaw puzzle that once fit together but have since been scattered across the floor, just like what happened when you didn't put away the 1,000-piece puzzle of the signing of the Declaration of Independence, and now John Hancock's quill and John Adams's right thigh are gone forever. What is less widely known is the reason for these chronic continental perambulations, these bumps and rasps of plate against plate. At which point, I must say it's a shame that respectable Christian theologians dispensed long ago with the idea of hell as a specific, corporeal, very hot and nasty place located deep underground and have replaced it with a flaccid metaphor along the lines of "hell is the spiritual desert in which one dwells if one turns away from God." As it happens, there *is* a raging inferno buried some 1,800 miles underground, an authentic hell in Earth, and it is none other than our planet's core. This pyred pit, this devils' spa and nail salon, is a ball of blazing metal roughly the size of Mars, 90 percent iron and the rest mostly nickel, and it burns at a temperature of 10,000 degrees Fahrenheit, nearly as hot as the surface of the sun. The core has been seething continuously since Earth coalesced, and with almost unmitigated brimstone, cooling by only 300 degrees over the past 4 billion years. Most of that heat is left over from the cauldron conditions of the early solar system, and from the inevitable transformation of potential energy into thermal energy that comes when gravity pulls a lot of scattered matter into a compact planetary ball. The rest is supplied by rich stores of unstable radioactive elements like uranium, thorium, and potassium, which in decaying release energy into their milieu, their terrestrial stew, and so keep stirring the pot. Earth is exceptionally blessed with radioactive material, and the feverish click, click, clickings of decomposing heavy atoms, along with the core's primal heat, explain why our planet is such a changeling, displaying more geologic verve and greater turnover of its dermal layer, its surface anatomy, than all the other planets of the solar system combined. Mars used to have a similar geologic profile, a blistering core driving large-scale upheaval — crackling crust, volcanoes spewing ash and gas. But Mars, being significantly smaller than Earth, had far less internalized heat and fissionable goods to begin with, and its furnace went cold a billion years ago, leaving the planet a relatively indolent world, its worn, pitted face pretty much set in its ways. The Earth, by contrast, is a mad master of plastic surgery, recurrent patient and outré doctor bundled as one. Do you think India looks good down here, cheek by jowl with Madagascar? No? Then how about sutured to China? And Australia: Better down here, blended together with Antarctica, or as a standalone flower of the Indian Ocean? Or maybe you'd prefer if we slid it northward and into Japan?

The surgery won't stop because Earth is a giant heat engine, and hot things are forever struggling to cool themselves down. This is the most telling way to think about our planet, said David Bercovici, a professor of geophysics at Yale University: as a great hot ball trying to shed thermal energy into space. After all, the second law of thermodynamics demands the transfer. Heat must travel from a spot of relative warmth to one of relative cold. The core of Earth is almost 6000 degrees Celsius. The space through which Earth is hurtling is about −270 Celsius. Ergo, the core keeps shrugging off its heat, Get out of here already, out into the frigid sinkhole of space, where you belong. Ah, but it's not always easy to play by the rules. Not only do the subterranean supplies of uranium and thorium keep pumping heat into the mix, but as thermal currents travel from the core through the thicket of inner earth, they must pass through thousands of miles of densely packed rock and metal and putty and pudding, and up through the thin, brittle, insulating crust, and all the while never knowing how the substrate they're seeking to penetrate will react — will it shrink, collapse, crack, balloon outward? It's a real challenge. Yet this is what our world is about: there is heat inside, and it wants to get out.

"It's the same as with your cup of coffee," said Bercovici. "Everything is trying to get into equilibrium with the great, cold, vacuous space. And in the process of cooling off and becoming cold and undrinkable, it does all kinds of cool things."

What sort of cool things might heat transfer bring? Let's slit the world open and look from within.

As I'll discuss more fully in the next chapter, on astronomy, Earth condensed about 4.5 billion years ago from the ring of rock and dirt that remained after the formation of the sun, itself the result of a large gaseous cloud heeding the gravitational call to compaction. Earth and the other planets formed quickly by celestial standards, accreting their mass and assuming their spheroid shape (the expected geometry when every part of an object's surface is pulled evenly toward the center by gravity) in as little as 10 million to 35 million years. Those early days were hard days, lawless days. The interplanetary skies were littered with comets, asteroids, and other extrastellar trash, and orbital paths were still in violent dispute. Some 50 million years after the birth of the solar system, Earth collided with a planet approximately half its size, to spectacular, double-barreled effect. A portion of that doomed planet's mass was absorbed into our own, for a net weight gain of 10 percent. At the same time, a chunk of the original Earth was knocked free in the crash, a prize piece that became our incidental, parthenogenic daughter and sole satellite, the moon.

The newly enlarged edition of Earth then began to assume its current configuration. Denser materials like iron and nickel were tugged at most strongly by the planet's gravitational field and gradually migrated toward the center. Lighter elements, including oxygen and silicon, felt less of a pull and formed intermediate and outer layers. And this, in rough cut, is the Earth we have today: an orb composed of a ridiculously dense metal pit surrounded by comparatively lighter, fluffier layers and topped off with a crispy outer crust. But this is no finished dessert. This is not the end of the meal. Great pressure and radioactivity keep the fires burning; and when the chef is disgruntled, everyone else feels the heat.

For those of us who live at or around ground level — and that includes all known life forms, for even the deep-ocean dwellers puddling around hydrothermal vents subsist well within the confines of the crust — it's hard to appreciate the power of pressure. Humans put up with atmospheric pressures; but though Earth's atmosphere is thick, as these things go, extending upward for 50 miles and more, and though we at sea level navigate through its bottommost, heaviest stratum, still it is a relatively flimsy load: only 14.7 pounds of air bear down on any square inch of us. But inside Earth, things get weighty fast. Each layer is a solid, or some equivalent of a solid, and every successive layer has to hold up under the weight of all the solid layers above it. Penetrate eighteen miles down, and you're talking about an average pressure of 150,000 pounds per square inch. Go two hundred miles in, and it's up to 1.5 million pounds per square inch. By the time you reach the innermost core, you encounter crushing loads of 50 million pounds per square inch, or some 3.5 million times the pressure of air.

Earth's core, its compacted ball of iron, nickel, and other chunky elements, is really a ball within a ball, an inner zone the size of the moon — about 2,600 kilometers across — surrounded by an outer core as wide as Mars. The inner core is where we have our 10,000-degree-Fahrenheit Hadean heat. That would be more than hot enough to melt iron under most circumstances, including on the similarly searing face of the sun. But great pressure makes hay of ordinary chemical change and packs iron atoms into such tight rows that they can't get up and flow. As a result, the inner core is a solid, something akin to a huge crystal ball of iron.

In the outer core, pressures are a bit more relaxed, and so, too, are the resident ingredients. The outer core, like the inner, consists primarily of iron, but here it slips around as a liquid. That fluidity has one particularly welcome spinoff effect, which helps make Earth hospitable to life. As the molten metal of the outer core glides around the solid iron of the inner, the motions generate Earth's magnetic fields, which could well be called magnetic shields. Extending outward into space for thousands of miles, the magnetic fields help to deflect much of the solar wind, the crackling cataract of high-energy particles that streams nonstop from the surface of the sun, and that would, if left unchecked, strip away at our atmosphere as surely as turpentine does paint. Terrestrial magnetism then colludes with the cosseted atmosphere to defend the planet's surface against the sun's most dangerous light. Together air and magnetic fields scatter most solar X-rays, cosmic rays, and gamma rays before the radiation can reach us and tatter our cells and genes.

Magnetic fields also infuse the world with a sense of place, an inherent cartography of north and south, and many creatures are thought to navigate by tapping into terrestrial magnetism: pigeons, sparrows, bobolinks, humpback whales, salmon, spiny lobsters, loggerhead turtles, monarch butterflies, newts, the Central Australian Bushwalking Club. Then there are those of us with absolutely no sense of direction, whose idea of using a compass amounts to handing it over to a park ranger in exchange for a helicopter lift.

The core, inner and outer together, accounts for only about a sixth of the volume of Earth, but a third of its mass. It is defined by its heaviness, the density of its components. The most formidable atoms of which Earth is formed, those with the greatest number of protons and neutrons, have been lured inward by gravity, and in their pigheaded procession toward the midpoint have pushed aside slimmer players that stood in their way. The concentration of iron, nickel, and their atomic ilk in the core, to the near exclusion of lighter elements, makes for an unmistakable boundary between core and noncore. When you move outward from the core and into the adjacent layer of Earth-meat, the mantle, the difference in density is as extreme as it is between that of the ground we stand on and the sky above it.

Most of Earth's girth is taken up by the mantle, a word that comes from the German term for "cloak," as the mantle cloaks the core. And though the cloak is much less dense than the core, do not mistake it for gossamer. It is solid, rock-solid, a vast and varied mosaic of metals and silicates — materials built mainly of chains of silicon and oxygen, which pretty much covers the whole terrain we call stone. One of the misconceptions that people have about the mantle is that it is molten, a big vat of melted rock sloshing around underground like the molten lava bubbling from the mouth of a Hawaiian volcano. In fact, while much of the mantle is close to its melting point, particularly in the regions closest to the core, very little is truly liquid. Instead, the mantle is more like Silly Putty, a toy that more than one geologist keeps on hand for demonstration purposes, and to make funny imprints from the newspaper when they're bored. Like Silly Putty, the mantle is solid but springy, almost squishy, and it can move, and it does all the time. "Think of glaciers," David Bercovici suggests. "They are solid ice, and they move. They go very slowly, but they go." The mantle, too, is a very slow goer. It flows like a great sheet of rubbery rock around Earth's central core, at a rate of up to ten centimeters a year, slower than the speed of growing hair.

Above the mantle is the planet's outermost layer, the real cloak of Earth and the place we know best — the crust. On the one hand, "crust" seems like an unnecessarily lackluster and dismissive word for something that has fed and housed us so well. The entirety of life on Earth lives on or in the crust. The seven continents and 100,000-plus inhabited islands of the world are part of the crust. The oceans and the floors they lie on are part of the crust. The beds from which we extract oil, natural gas, and coal are part of the crust. In crust we trust, and it has ever been thus.

On the other hand, the crust is *very* thin. It is a measly, miserly submorsel of Earth, accounting for less than one-half of 1 percent of Earth's mass and 1 percent of its volume. If you were in prison and somebody threw you a crust of bread with the same proportion of the original loaf as the earth's crust is of Earth, its breadth would be half a millimeter, barely thicker than a couple of eyelashes. The planetary crust is of such insignificant width compared to the entire sphere that if you reduced Earth to the size of a basketball, the skin on it would be much smoother than that of a basketball, closer to that on a bowling ball. All the nervy peaks and plunging vales we take such pride in conquering would vanish, leveled by force of contrast with the bulk of the mantle and core.

One fair way to think about the crust is as a layer of ice on a lake. The ice floats because it is lighter and less dense than the water below it, and it is crispily crystallized because it has been chilled by the winter air above. So, too, is the crust composed of relatively light rocks buoyed atop the condensed superputty of the mantle. The crust is also the coolest part of Earth, and thus is brittle and prone to fracturing. And just as the ice on a lake is thicker in some parts than others, which is why it's a very stupid thing to try driving your Volkswagen Beetle across it no matter what your cousin Jeb tells you, so the width of the crust varies considerably, from a thin point of three miles for the ocean floor beneath Hawaii, to a thickness of about forty-three miles for the crust of the Himalayan plateau. In general, continental crust is about six or seven times thicker than oceanic crust; and though the benthic realms of the ocean floor have a spooky, primordial mystique about them, the place where you might expect to find a few surviving trilobites, the lost island of Atlantis, or at least the original cast and crew of *The Love Boat*, in fact much of the seabed is quite young, hundreds of millions to billions of years younger than the dry land on which we stand. Which brings us to the sublime theory of plate tectonics, a fundamental organizing principle of geology and one of the grand discoveries of the twentieth century. It also returns us to the image of the very hot object that wants to cool down — a cup of coffee, a bowl of porridge, a planet with an iron smelter of a core. No matter: the patterns that arise when heat bubbles up from below will resemble one another through whatever substrate they flow.

The idea that landmasses migrate slowly around the planet is not a new one. As maps improved, scientists and others couldn't help but puzzle over the puzzle-piece appearance of the continents. Incisively fusing evidence from fossils, rock deposits, and glacial striation patterns around the world, the German geologist and meteorologist Alfred Wegener in 1912 published his visionary "continental drift" hypothesis. Wegener proposed that 200 million years ago all the continents were lumped together in a giant landmass he called Pangaea, meaning "All Earth," and that somehow Pangaea had broken into pieces and the shards had drifted apart. Incidentally fusing the name of a fictional detective with that of the character's creator, the English geologist Arthur Holmes soon afterward suggested a possible mechanism for Wegener's continental peregrinations. Holmes, who studied physics and geology at what is now the Imperial College in London, conjectured that ongoing radioactive decay in the earth could be helping to generate giant heat currents that would convect up to the surface like soup cooking on a stove. Not until after World War II, however, did scientists gather empirical evidence that a steady spreading of the sea floors, fueled by chthonic radioactivity, was driving continental drift; and not until the 1960s did geologists pull all the pieces together into a grand unified theory of how the earth churns. The theory of plate tectonics is a bona fide theory, too, a comfortably capacious conceptual framework that explains an array of disparate findings, that grows stronger and sturdier with the steady accretion of new data, and that can be used to formulate and test all sorts of novel, unobvious hypotheses about how Earth behaves. Scientists may not yet be able to predict earthquakes or volcanic eruptions with anything near the precision that they, we, and the insurance industry would like, but they can make actuarial predictions about where and over what time period the bigger quakes are bound to happen.

The term "tectonic shift" has filtered into popular usage and competes with "quantum leap" to suggest a really big, generally constructive, but possibly risky change, all of which are appropriate nuances: "tectonic" comes from "tekton," the Greek word for "builder." Plate tectonics is the theory of how the shifting plates of the earth build the great bulk of our surroundings. And construction sites can be dangerous places — why do you think the workers wear hardhats, carry metal lunch boxes, and know how to whistle? Earth's tectonic plates build, wrench apart, gerrymander entire nations on a lark, but they are not what they seem. You can't look at the globe and know where the plates are. They aren't defined by the shapes of the continents or where land meets sea. In fact, tracing the borders of the earth's plates is a tricky, sometimes fractious task. By general consensus, there are seven to ten large or "major" plates, and twenty-five to thirty minor ones. The precise head count counts far less than how the plates move, where they are headed, and what happens when two collide.

What, then, are the tectonic plates? Contrary to common misunderstanding, they are not simply broken pieces of the earth's crust, although the crustal rock is usually fractured along the boundary line between one plate and the next. But the plates extend deeper than the crust, into the upper part of the mantle. Each is about fifty miles thick, although, like the crust itself, the plates vary in width and in density, too. The plates bearing the continents are relatively thick and light, while those scalloped with ocean basins are thin and dense. The plates are defined as much by their motions as anything else. They are the segments of the outer earth that glide around as reasonably cohesive units. The upper part of each mobile plate, the crusty part, is brittle and prone to cracking and crumpling. The nether portion, in the mantle, is hotter and more plastic, more likely to yield when pressed. All the plates are sliding over, or in some cases with, the more viscous lower mantle beneath, at an average clip of one to ten centimeters per year — about the rate at which your fingernails grow. That may be snail-paced by human standards, but truly nail-biting by geologic ones. In a million years, a plate migrates some 30 miles. Give a plate 100 million years, and it will have globetrotted 3,000 miles, nearly the distance between New York and London.

Animating the plates is Earth's ceaseless effort to dispel its suffocating heat. The planet has a few cooling techniques at its disposal. It radiates a small amount through conduction, in which fast-moving atoms and molecules pawn some of their excess energy off onto slower-moving atoms and molecules that abut them — the same process that quickly heats your metal spoon when you stick it in a hot cup of coffee. Earth expels another modest degree of thermal energy by straightforward mechanical venting — volcanic explosions, geysers, and related geo-belches. Mostly, however, Earth relies on the conveyor belt method of heat dispersal, convection. Convection has the benefit of cooling not only by pushing hot things outward, but by pulling cooler items closer. The convective currents that course through our world are complicated and difficult to track, rather like large-scale weather patterns in the atmosphere, but here in crude cut is what happens. Heat flows from the iron core and into the rock of the lower mantle. As that boundary rock heats up, it expands and becomes less dense, and, just as hot air rises, so the heated, expanded rock starts to rise through the cooler mantle rock above it. The higher it manages to climb, the less pressure bears down upon it, and the softer it can become; and the more buttery it grows, the better it flows, which further eases its crustward cruise. At some point, however, another little technicality of physics intervenes, the flip side of the principle that sent the rocky mass bubbling upward to begin with. In rising, the rock dispenses its heat into its environs, and as it cools it gradually reverts to its former state of density. Finally, the bleb of stone has no choice, it is too heavy relative to the matrix around it, and it begins to sink, as stones predictably, platitudinously do. Down, down, back toward the hotter core, where the rocky mass can pick up more heat and start its yearning journey once more. This, then, is the basic convective cycle at work inside Earth. Hot rocks expand, rise, cool, contract, and descend; let's take a deep breath, and try that again. Some of these convective currents may eddy around near the border of the core, others convulse in larger swags across vast swaths of the mantle. And a few manage to fight their way to the surface and gurgle into the oceans, right where the crust is thin and the Earth's seams ever so slatternly gap.

Among the many lines of research that culminated in the theory of plate tectonics, some of the most important came from studies of the sea floor in the 1950s. That enterprise supplied a braided array of surprises. For one thing, there were long ridges of undersea mountains, the most prominent running down the middle of the Atlantic and Indian oceans and rising 3,000 meters or more above the sea floor; and there were undersea trenches, plunging 2,000 meters or more below that floor. For another, the rocks on the sea floor were outrageously youthful, 180 million years old at most, compared to terrestrial rock samples that date back billions of years. The youngest of those spritely seabed rocks were found closest to the midocean mountain ridges, with the ages steadily mounting as you moved farther from the mountains and up to the rim of the midocean trenches. Finally, the sea floor proved to be remarkably tidy and light on sediment, considering how long it had been subject to a steady drizzle of debris from the land above — posthumous plant and animal parts, sand, pebbles, mud, bones, shells, Naugahyde barstools, and 3,000 copies of Grand Funk Railroad's *We're an American Band,* still in their unopened jewel cases. It was as if the ocean bottom were being continuously scrubbed, vacuumed, and mercifully auctioned off on eBay.

Plate tectonics helped solve the riddle of the callow deep. The convecting loops carry roasted rock upward toward the carapace. Some of that hot young rock penetrates to the surface by welling up through the ocean ridges, where it emerges in the semisolid rock format called magma. That magma pushes the ocean floor apart, bobbling the oceanic plates on either side and plowing cooler, less young rock outward. Eventually, the cold leading edge of the spreading sea floor runs into other fissures in the crust, the deep ocean trenches, where it is sucked back, or subducted, into the mantle. In the mantle's mulching maw, the rock is smashed, pulverized, refitted, and pasteurized, so that whatever part of it might manage to emerge through a crustal ridge, to see the sea floor by the seashore once more, will do so as brand-new rock. Poseidon's conveyor belt never stops rattling along, and in Earth's hoary history the ocean basins, our crustal low points, have been recycled dozens of times. Not so the continents. Because continental rock is relatively light, it floats above the subduction zone of the trenches, getting pushed and pulled and battered without being routinely sucked into the mantle. Continental landmasses have changed their contours and allegiances repeatedly, as we said, but many of their rocks have remained above the molten fray for a billion years or more.

Everywhere the upwelling of hot rock keeps the plates in constant motion, and the moving plates in turn recast the crustal playhouse on which life gamely mouths its lines. The spreading of the sea floor at the midocean ridges drives some plates and the cargo they bear away from one another: such tectonic divergence is pushing North America and Eurasia in opposite directions, and widening the Atlantic Ocean by some five centimeters per year. Other plates are ramming into each other, as awkwardly and irritably as two pedestrians colliding on the sidewalk: You go this way, no, I'll go that way, whoops, now we're both going the same way again, that won't work. Maybe I'll just try to duck under your legs and get this farce over with. When a thick continental plate rubs up against a thin oceanic plate, the thinner plate does indeed start diving under the higher plate, making one of those subduction zones that return old sea floor to the mantle and seriously unsettling whatever landforms lie above it in the process: raising a string of volcanoes and outfitting them with explosive magma chambers, for example, or crumpling coastlines into high-altitude peaks best suited for llamas, kings with lots of slaves, and tourists with lots of medical insurance. The Cascade Mountains of the Northwest — home to Mount Saint Helens — and the Andes Mountains of South America both exemplify what happens when oceanic and continental plates collide.

If the converging plates both carry continents, the landmasses will be smashed together in slow motion, the leading edges buckling upward as the continents are forced into uneasy, captious alliance. Mid-continent mountain chains often reveal where once separate landforms were compressed together by converging continental plates. The Himalayas, for example, began clawing their way upward about 45 million years ago, when the plate bearing the Indian subcontinent collided with the rest of Asia. In Europe, the Alps delineate where the Italian peninsula, riding on the African plate, slammed into what are today Germany and France at about the same time, a churlish merger that two world wars, a common currency, and the frequent consumption of each other's pastries have not entirely placated.

Tectonic encounters are not always head-on collisions. Sometimes a pair of plates traveling in opposite directions merely scrape past each other, or try to. If the glancing encounter turns out to be a tight squeeze, parts of the plates will stick together, especially at their crispy ragged upper crusts. The plates underneath may insist on continuing in their respective, contradictory directions, but the rocks along the tacky upper boundaries are trapped in place. They become stressed and strained and resort to all sorts of tricks — therapy, yoga, renaming themselves "Gibraltar." But the pressure keeps building, and finally the strained rock surfaces snap, lurching away from each other in a seismic spasm. "Seismic" comes from the Greek word for "shaking," and sudden slippages along Earth's fault lines — fissures in the crust where underlying plate motions force rock to scrape against rock — are what shake and break the ground in a quake, as the pent-up energy stored in the long-suffering rocks is freed to spread outward in waves.

The most famous of these parlous plate boundaries is the San Andreas Fault in California, where the Pacific plate is crawling northward relative to the North American plate, and their stony interfaces alternately grip and slip, usually in incremental jerks, occasionally by harrowing heaves of several meters at once. In the catastrophic San Francisco earthquake of 1906, peak displacement was almost twenty feet near Olema, California. The chronic grinding and sliding of plates tend to fracture boundary rocks in many directions and along multiple planes, down an estimated six miles deep in the case of the 1906 quake. As a result, major fault lines like the San Andreas are not single slashes in the crust but crisscrossing thickets of cracked rock slabs, which sometimes absorb the querulous motions of the plates and sometimes recoil from the effort and splinter some more. The difficulty of determining the relative resilience of any strand in a fault line's sticky thicket explains the considerable challenge of predicting when the next earthquake will strike, and exactly how bad it will be.

The prodigious upwelling of the mantle has done more for Earth's crust than hammering it *ad infinitum* with magmatic rancor. In addition to making the world's seabeds, the convecting forces of inner earth supplied the water that lies on them. Earth, of course, is awash in water. There are 326 million trillion gallons of it, enough to cover three-quarters of the planet's surface with flowing oceans that average 2.5 miles deep. Liquid water is essential to life as we know it, and none of our sibling planets can claim anything like our aqueous bounty. The precise sequence of events that pinned the bright blue rippling ribbon on Earth's lapel is still open to debate, but most scientists concur that it was likely a mix of the astral and the retentive. Liquid water may be rare in the solar system (and, as far as we can tell, in the universe generally), but H_2O in its other states is not. Comets abound at the fringes of our solar system, and we can in clean conscience refer to them as "dirty snowballs." A comet is nothing but an orbiting chunk of ice and dust maybe ten miles across; and the dramatic tail that cries "comet" so clearly that we recognize its image in the thousand-year-old Bayeux tapestry is a humble puff of steam, the surface ice boiling off as the speckled projectile careens close to the sun. Early in the evolution of the solar system, it seems, wild swarms of comets were drawn in from the exurbs by the highly credible gravitational pull of Jupiter, the giant of the planetary litter. A sizable number of those comets either forgot to bring their portable GPS device, or found it wasn't working because it hadn't been invented yet, but in any case they ended up overshooting their target by a few hundred million miles and crashing into Earth instead. Earth was still young, and so hot to the touch that much of the cometary water rapidly vaporized back into space; but some seeped into the depths of the young Earth, where it was bound up into the rock, greatly amplifying whatever water stores our world had from the start. Beginning about 4 billion years ago, volcanic eruptions steadily loosened the subterranean water from its mineral crypt and spewed it back out as steam, up to a planetary surface grown clement enough for the water to rally its forces, seize the initiative, and initiate the seas. Earth's crust had cooled down, and the swirling molten iron at its core had begun generating the magnetic fields that help deflect the searing solar wind. So shielded, the vast clouds of volcanic reflux were not peeled off into space, but instead hovered above the ground, gathering and glowering portentously, as storm clouds do. They gathered until there was no room for more. The skies were supersaturated out to the limits of gravity's grip, in a nimbus of high humidity, and the water vapor had no choice but to condense out and fall to Earth as rain. The rains fell in relentless torrents of Noachian proportions, only bigger and longer, and there were no giraffes and zebras huddled on a wood boat, wishing they were back at the circus. For tens of thousands, hundreds of thousands of years, the antediluvian deluge fell, filling the dimples in Earth's newly caramelized silicate skin, filling them to the brim. Yet while this downpour may sound like an excessive rainy season, even for Seattle, by the timetable of Earth's history the building of our naval resources barely lasted a sneeze. "The geological record of sedimentary rocks, formed in the presence of liquid water," the geologist Robert Kandel has written, "proves that the oceans have existed for three billion years, maybe even four," and at a volume pretty much like what we have today. In other words, no sooner had Earth's crust managed to cool enough for discernible depressions to form than the skies had deposited in them the maximum allowable in liquid assets, our storied galleons of 326 million trillion gallons, enough water to fill a string of bathtubs that would reach from here to the sun and back again 5 million times. Just be sure to bring your own towel, shampoo, and a breathable atmosphere, too.

Oh, yes, a breathable atmosphere. How easily we overlook what we need the most. You can live without water for three days, even a week to ten days if you're well hydrated at the start and you stay in a cool, shady spot. But if you stop breathing, you can die within minutes. Our obligingly inspirable air may seem simpler and less substantive than water, more insipid and inadvertent and with a more pronounced tendency to stare vacantly into space, yet appearances can be deceiving, especially the ones that you can't see. In fact, Earth's atmosphere is a richer and more complex resource than is our water, and it has taken comparatively longer to evolve into the specific mix that we inhale at a rate of 2 gallons every minute or so, for a total of about 3,000 gallons per day—enough to fill 100 bathtubs not currently en route to the sun. Because gas has mass, a pound for every 100 gallons, we pull in and blow off about 30 pounds of personal space daily.

The atmosphere is really an extension of Earth, a geopolitical player right up there with the core, mantle, and crust. Like so much else about our planet, the air was born on the inside and then brusquely turned inside out. From the moment Earth managed to cohere into a reputable sphere, it began outgassing the rudiments of an atmosphere, releasing plumes of hot vapors that had gotten trapped in its rocks during the natal melee. The first atmosphere was mostly hydrogen and helium, and it wasn't long for this world. Earth lacked sufficient gravitational mass to hold on to such lightweight gases, and its young core had yet to settle into the two-part inventor of a magnetic buffer against the solar wind. Before Earth had reached the half-billion-year mark, its primordial atmosphere had drifted or been abraded away. And while there are traces of helium in our skies today — a tiny fraction of 1 percent of the total atmosphere — there is virtually no free hydrogen gas to be found. Terrestrial hydrogen is in bondage with other elements — with oxygen in water, with carbon and nitrogen in the chains of our genes and proteins. If you want hydrogen straight up to put to work in some way, say as hydrogen fuel to power high-concept hydrogen cars, you must wrest it away from its molecular setting, and that takes energy, too.

Earth's second atmosphere would not be so easily spooked as the first. The crust cooled, and pertussive volcanoes loosened other volatiles from subterranean stone, outgassing water vapor, nitrogen, carbon dioxide, and ammonia at a furious pace, until the skies held one hundred times more gas than they do today. From this toxic, lofted broth, water vapor condensed as rain, and that was the start of the seas we see, and the earliest inklings of the air we need. Once poured, the oceans began absorbing some of the other gases in the atmosphere, dissolving with particular relish the carbon dioxide at its surface and transforming it into a seltzered froth. Ocean currents stirred carbon dioxide microbubbles wide and deep, until fully half the carbon dioxide from the atmosphere had been sucked into the sea. Bubbles: Everybody loves bubbles! Blowing them, drinking them, bathing in them, bursting someone else's. Bubbles are like puppies, always bouncy and happy and ready to play. What a shame if there were nobody around to grab the leash and take all those lively carbon-based bubbles out for a romp, running and running until somebody, somewhere, stops and takes a deep breath.

We don't know how life began on this planet. We don't know where it started — in surface waters littered with sunlight, or on the black ocean bottom by a piping hot vent, sheltered in a calm crescent of clay or slapped sentient by intertidal spray. We don't know when life started; estimates range from 3.2 billion to 3.8 billion years ago. We don't know what the very earliest life forms were like. But we do know that once life got started on this restless, bibulous Goldilocks planet, it did as Goldilocks would, upending everything in sight until the place looked and felt and smelled like home.

The impact of life on Earth was dramatic, a tectonic shift of its own, and nothing illustrates that impact better than the things life did to the air. The atmosphere in which life arose, that special blend of next-generation outgassings from our generously ulcerous underground, was wondrous for its time, and likely offered the ideal setting for the chemistry of life to get its first hesitant footing, but it was not the sort of air the vast majority of modern organisms would describe as "fresh." Most notably, the atmosphere had no free oxygen in it. Yes, there were oxygen atoms bobbing about with their hydrogen earpieces, in the ambient water vapor, but of the paired, pure oxygen, the O_2 gas we need to breathe, the air was almost entirely bare. Today, the atmosphere is about 20 percent O_2. Who put the pairs there? Our self-sacrificing ancestors, the cyanobacteria: large, floating mats of sun-eating microbes, ur-solar cells that made sweetness from light. Cyanobacteria, also called blue-green algae, were among the earliest known life forms, and a great success story. They were probably the first to master the art of photosynthesis, the stepwise transformation of solar energy, water, and carbon into sugar, the all-purpose cell food. Sunlight fell in abundance, and water — well, they *were* aquatic. And for the carbon source, they had the bubbles, the carbon dioxide that fizzled into the water from the air, and the matted flats of cyanobacteria greedily gulped it in. From the CO_2, they took the Cs they needed to bake their carbohydrates, their daily bread, and excreted the parts they couldn't use, the oxygen couplets, the mighty O_2s. Yet the air remained long unruffled, as all the oxygen waste spilling from Earth's booming archaeobacterial farms went instead, quite literally, to rust. The oceans were rich in iron — dissolved in the water or veining submerged rocks — and iron has a great affinity for oxygen. For the first billion-plus years of photosynthetic activity, oceanic iron handily sopped up the oxygen, and to this day most of the free O_2 ever made in Earth's history remains locked up in ancient reservoirs of red, rusted rock.

Still, life kept up the pace, and the bacterial mats spread, until, beginning about 2 billion years ago, the sea's supply of exposed iron was oxidized, pig-sick of O_2, and the excess oxygen started filtering into the atmosphere. As it built up, some of it occasionally reacted with itself to form O_3, an ozone layer, which in turn helped block out ultraviolet waves from the sun. Life below was growing steadily better for growing, in number, in kind, and in setting. The ozone shield would allow life to colonize the land without fear of frying, while the mounting count of oxygen duets in the air would spark the great aerobic revolution.

Cyanobacteria are still around today, numbering some 7,500 species, and many of those strains are, as their ancestors were, anaerobic, performing all their daily tasks with no need of oxygen. Indeed, exposure to oxygen will kill them, as it does other exclusively anaerobic microbes, including some of the symbiotic bacteria that live in our intestines, and other, less genial germs that cause tetanus and botulism. An anaerobically styled metabolism has its uses: it allows microbes to survive where nothing else can, and in our own bodies it gives muscle cells a chance to flex for short bursts of intense activity when our blood can't deliver the requisite oxygen in a timely fashion. Yet oxygen is an excellent fuel if you know how to burn it, and aerobically powered cells run far longer and more efficiently than their anaerobic counterparts. Aerobic strains of bacteria can divide thirty to fifty times more quickly than anaerobic ones. And while you can sprint for just a minute or two on the fruits of anaerobic metabolism alone, if you slow to a measured pace that gives your circulatory system a chance to supply oxygen as needed you can keep running for hours, the whole day if you're training for the Olympics, or owe a lot of money to an unofficial lending source in New Jersey.

Some time around 1.5 to 2 billion years ago, as oxygen concentrations climbed toward 1 percent of the atmosphere's gaseous mix, the first aerobic microbes arose, the first unicellular organisms that could exploit free-floating oxygen to power their internal operations. Dividing at a quickened pace, the oxygenic microbes began their sometimes rocky climb to dominance. They'd crowd out the anaerobes or subsume them beneath their more busily bulging blankets, only to begin exhausting the oxygen that their blue-green rivals supplied. The aerobics would crash, and the anaerobics revive, and oxygen levels rise again. The benefits of a dual survival plan, of burning oxygen when possible and switching to an alternative, oxygen-free strategy when necessary, must have occurred to one of these ancestral life forms, occurred in the sense of, yes, it happened, it occurred, and it was good. The first eukaryotic cells, the first cells to have their genetic material cloaked in a nucleus and to be otherwise well organized and compartmentalized compared to bacterial cells, are thought to be the result of an archaic merger between the distinct cell types. It may have been an accident, it may have been wolfish engulfment with a fairy-tale ending, we don't know, but the molecular and metabolic makeup of our cells, of all eukaryotic cells, suggests that early on some sort of large anaerobic cell — not a blue-green algae but an anaerobe that ate other cells rather than synthesizing its food from scratch — either fused with, swallowed, or was infected by a smaller aerobic cell. Rather than be digested down for spare parts, the smaller cell survived in the cytoplasmic sanctuary of the larger cell, and therein arose one of the world's first great symbiotic partnerships. The larger cell protected the smaller cell and fed it anaerobically whenever oxygen proved scarce, while the smaller cell powered its patron through aerobic respiration whenever oxygen molecules diffused into the amalgamated microbe's gelatinous interior and aroused the aerobe's interest. These early switch-hitting cells were a bit clumsy, and they must have stumbled down a number of dead ends and blind alleys as they struggled to sort out the business of cell division complicated by the need to replicate and properly parse two cellular species rather than one. But their newfound metabolic plasticity and chemical deftness lent them sufficient advantage that they thrived despite taking longer to divide than purely aerobic microbes.

Today we see the purest expression of this ancient alliance in yeast cells, considered the most "primitive" of eukaryotic cells but no less worth toasting for that. Yeast cells have their distinctive aerobic and anaerobic phases: the first phase begins to bubble your beer, the second ferments it. But all eukaryotic cells carry living proof of the primal alliance. Look at any one of your body cells under a powerful microscope, and find the mitochondria, the striped, sausage-shaped bodies where oxygen is burned and food molecules are transformed into energy packets, to be stored or spent as needed. Those mitochondria are the descendants of formerly free-swimming cells; and though they have long since forsaken the means to survive on their own, mitochondria keep pieces of past freedom in their small stash of genes. Mitochondrial DNA is distinct from the much larger cell genome stored in the nucleus, and its limited number of genes encode proteins devoted mostly to aerobic affairs and energy production. No other component of our large, crowded cells has even this modest measure of genomic autonomy. The mitochondrial exception was written into the original eukaryotic compromise, and through more than a billion years of evolution it has never been broken.

There were other early instances of a cellular pooling of talents. Today's plant cells are thought to be the result of an ancient encounter between a cyanobacterial cell, with its priceless sun-eating chemistry set, and an aerobic cell that could make good on the oxygen wealth in the air. True to the terms of that paleocoupling, modern plants live a Jekyll-and-Hyde sort of life. During the day, when solar energy galvanizes their photosynthetic machinery, plants breathe in carbon dioxide, make their sugars, and exhale oxygen gas, in a manner reminiscent of cyanobacteria. But at night, plants take small amounts of that oxygen back, reabsorbing the gas through diffusion and using it to help transport their homemade food plantwide.

For all the seesaw cycling between aerobic and anaerobic life, the level of atmospheric oxygen gradually mounted until about 400 million years ago, when it reached a concentration much like what we see today, a fifth of the total ether — though there have been fluctuations up and down ever since. Scientists have cited the surging supply of oxygen as a likely stimulant for a number of evolutionary seismic shakes. One was the advent of multicellular life around 700 million years ago, when heretofore separate eukaryotic cells began banding together into interdependent clans and taking up specialties — I'll be the mouthparts if you'll serve as the gut tube. Another was the so-called Cambrian explosion of 530 million years ago, the dramatic diversification of multicellular life into a bona fide bestiary, the fete of faunal body plans that included the ancestors of all the major animal groups alive today. Some researchers also attribute the formidably proportioned arthropods of the Carboniferous period, roughly 300 million years ago, when dragonflies had wings like falcons and scorpions were the size of skunks, to a sharp spike in atmospheric oxygen, the result of an exponential growth of vascular plants. Even today, regions of comparatively high oxygen concentrations are often home to unusually large invertebrate species. The biggest jellyfish and marine worms are found in the coldest, most oxygen-rich waters of the ocean. The correlation between giantism and oxygen is not absolute, however; and, as far as I can tell, urban insects that inhabit poorly ventilated spaces like cupboards and basements seem perfectly capable of turning Goliath on the spice of spite alone.

The ceaseless give and take between bio and geo doesn't stop at oxygen. Carbon is cycled in great, intersecting loops through water, air, mud, body plans living and dead, now drifting into the atmosphere as carbon dioxide gas, now sinking into sediment as rotting gymnosperm forests. Calcium snakes through rocks, water, seashells, our cells. Iron and other trace metals play pivotal roles in both the private biochemistry of the body and the public geochemistry of the oceans, and the amount monopolized by one party at a given moment affects the rhythms and possibilities of the other.

We live on a Goldilocks world, taking the jackpot trail through the solar system. Go one planet closer in, to Venus, and temperatures average 900 degrees Fahrenheit. Jump out a groove, to Mars, and it's −75 degrees. Earth is just right for life, and life has clung to its skin for more than 3 billion years, if sometimes just by the skin of life's teeth: 99 percent of all species that have ever lived have gone extinct. We humans may be exceptionally highhanded and ham-fisted in our Earthly transactions, but Earth and its life are much bigger than we, and they'll carry on whether we do or not. Maybe we need to get away from it all, take the ultimate off-road vacation to a celestial location. It's time for the populist stage of the Space Age, and the family-budget space flights we've been expecting since NASA gave us Tang. Everybody deserves a chance to experience the great awakening that clearly comes with extraterritoriality. Astronauts have attested to it time and again, the transformative moment when they first looked down on the oneness of bright blue marble Earth, their only home, and Earth looked back and said, *I know.*

9

Astronomy

Heavenly Creatures

FOR MANY OF US, the most memorable books, poems, and cautionary doggerel of childhood all had something to do with astronomy. We learned to wish on the star light, star bright, of the first star we saw that night, and we were incidentally saddled with lasting confusion over the precise semantic distinction between "I wish I may" and "I wish I might." We were asked whether we'd like to "swing on a star" and be better off than we are, or would we settle for life as a dirty, illiterate farm animal with disgraceful shoes. In the proxy of a rabbit wearing striped pajamas, we bade, "Goodnight, Moon," and goodnight, cow jumping over the moon, and goodnight, bears and chairs, stars and air, brush and mush, and, you, too, old lady, whispering "hush," and who are you, anyway, and how did you get into my great green room?

Even growing up in the Bronx, where any twinkling lights one might see overhead most likely belonged to a police helicopter, I had celestially themed dreams. In my favorite, I dreamed at age five that my family was vacationing in the country, and somebody called me to come look at the Milky Way; and when I ran outside and gazed upward, the heavens burst into tinkly music like that of a Mister Softee truck and drizzled me with milk. What a simple, joyful dream it was, and what luck that I wasn't a bed wetter!

We are all of us starstruck from the start, mesmerized by the spangled velvet of the nighttime sky, now longing to pull it close, like a mother, now shrinking beneath its inviolate diamond detachment. Soon we are able to pick out at least a few of the easier constellations — certainly the Big Dipper, maybe the Little Dipper, too, and boxy Orion with his bright belt and sword, and the five-star zigzag of Cassiopeia. We learn to distinguish between stars and planets by whether they twinkle or shine, for stars are so distant that they appear as mere points of light in the sky, and that light is easily bent and bobbled by turbulence in our atmosphere, while the planets are close enough that their radiance passes through air with scarcely a diversion or refraction, and so planets will bluntly, unwinkingly shine. Indeed, with an ordinary backyard telescope and under the right conditions, you can see the cheeky spheroid faces of our siblings in the solar system — Jupiter and its red spot, which is really a giant gaseous hurricane big enough to engulf three Earths and which has lasted for at least four hundred years; Saturn and its hallmark Hula-Hoops of ice, dust, and rock; tangerine Mars and moon-white Venus. But even our most powerful telescopes cannot resolve the disk of an extrasolar star, no matter how massive the star may be; all stars are too far away to be sized and analyzed as anything but points of light.

We stare and stare at the night, looking for something, anything, to make sense of the thundering silence — voiceover, pantomime, anagram, Vulcan mind-meld. Can't you just say something? Don't you hear us? Here we are! And as we stare, we see a streak of light, a wild platinum cat scratch piercing the mute tuxedo screen, and we're thrilled, each time, and filled again with goofy hope. A shooting star! I saw a shooting star! Did you? Well, just keep looking. You'll see one, too. Oh, we know they're not stars. They are meteoroids, space debris, the bits of interplanetary rock with which our solar system is littered; and though most of them are quite small, no bigger than a marble, they careen parabolically through space at such high speeds that when one of them hits Earth's atmosphere, the force of friction sets the rock ablaze, and Earth-bound viewers for thousands of miles around can watch the combusting rock bid us all a bright good-night.

With their tragicomic displays delivered in live-stream feed, meteors are especially easy for us modern humans to love and humanize, yet as Earth makes its squashed circle pilgrimage around the sun, the other stars and planets also appear to march across the nighttime sky. And the moon, as it wheels around Earth, swells and shrinks and swells again, not randomly, not like a yo-yo dieter, but in meticulous clockwork slices. The ancients missed not a trick or a tock. Like our nursery jingles and semiotic bunny board books, the earliest artifacts of civilization highlight our long-held fascination with the lights on high. Some 35,000 years ago, a sculptor-skywatcher living in what is now the Lebombo Mountains of southern Africa carved twenty-nine evenly spaced notches into a baboon bone, each groove likely representing a phase of the moon. Other artisans of the Pleistocene left behind similarly tooled eagle bones in sites not far from the famed Lascaux cave paintings in France. Ancient Chinese scholars engraved astronomy charts in bones and turtle shells, recording the paths of stars and planets and identifying hundreds of constellations. The dour megalithic monument of Stonehenge and the Mayan city of Palenque are thought to have served as astronomical observatories, their structures aligned to make dramatic use of the sun's light on the summer solstice, a sacred day in many cultures. For our seven-day week we can thank the ancient Babylonians and Greeks, who carefully observed the behavior of the sun, moon, and five quirky "stars" that we now know to be planets — the five planets that can be seen with the naked eye and that are so comparatively close to us they fairly glide across the sky, noticeably shifting their position against the stellar backdrop from one night to the next. (The word "planet," in fact, stems from the Greek word for "wanderer.") The seven heavenly standouts were named for the reigning deities of the era, and since every god must have its day, the names of the days followed suit. The Roman Empire and its Germanic outposts changed the names of the Greek gods while leaving the basic tenets of the pantheon intact; and though the Anglo-Saxon renderings of the deities can obscure some of the connections in English between the names of the days and of their heavenly projections, if you have even a smattering of familiarity with a Latinate language like Spanish or French, you can piece the little celestial-seminal puzzle together. Sunday is Sunday. Monday is Moonday. Tuesday in Spanish is *martes,* so Marsday. Wednesday, *miércoles,* or Mercuryday. Thursday is *jueves,* which, by Jove, is Jupiterday. Friday, or *viernes,* is Venusday. Saturday is Saturnday, my favorite day, a day for the unrestrained revelry of a saturnalia, or the protracted gloom of the hopelessly saturnine.

Throughout the ages, those who were wise in the ways of the skies were regarded as high priests and sages, petitioned for guidance on when to plant crops, woo a lover, launch a voyage, invade a country. In the predictable procession of the stars across the cosmic dais, people saw signs of divine intent, of a structure and certainty otherwise absent from their lives. Down on Earth, who could say if tomorrow would bring feast or famine, lockjaw or locusts? Up above, you knew that come the next spring, Virgo the Maid would appear in the southeastern sky. Humanity's fate was seen as bound up with the stars, a conviction yielding both the fantastical nostrums of astrology and the birth of the global economy. By setting their sights on the steady wink of Polaris, the star of the north, early traders could traverse the most pokerfaced seas and still find their way home in the dark.

Astronomers today may no longer be counted among the cultural clergy, and sometimes they complain about being comically misunderstood. "I don't do horoscopes, and I am not a failed astronaut," said Alex Filippenko, an astronomer at the University of California. Yet in the main, astronomers are among the most admired and beloved of scientists, and they know it and they like it. "We enjoy considerable public appreciation, and we get more than our fair share of press," said Chuck Steidel, a professor of astronomy at Caltech. "I'm always amazed at how I can go to my doctor or dentist, and they'll have a long list of questions for me when I arrive.

"Compared to something like high-energy physics," he added, "I'd say we have it really easy."

Astronomy is so easy to love. It is filled with outrageous magic that also happens to be true: novas and supernovas and pulsar stars that spin and click and are as thick as an atomic heart, as thick as Joyce's Muster Mark; and those thicker, darker collapsed star carcasses we call black holes, which are so dense that even light cannot escape their gravitational grip; and quasars, celestial furnaces at the edge of the known universe that are the size of stars but as luminous as entire galaxies; and theoretical plausibilities like extra dimensions beyond the four we know, or the creasing of space-time into shortcut "wormholes," which, if they exist, would be the equivalent of time-travel machines. Astronomy is about the heavens, the divinest of final frontiers and the presumed zip code of Ra, Vishnu, Zeus, Odin, Tezcatlipoca, Yahweh, Our Father Who Art In, and a host of other holy hosts; and that religious resonance markedly broadens the discipline's appeal, making it feel both cozier and more profound than it might otherwise. Astronomy also seems chaster than other sciences, purer of heart and freer of impurities, mutagens, teratogens, animal testing. Fairly or not, physics is associated with nuclear bombs and nuclear waste, chemistry with pesticides, biology with Frankenfood and designer-gene superbabies. But astronomers are like responsible ecotourists, squinting at the scenery through high-quality optical devices, taking nothing but images that may be computer-enhanced for public distribution, leaving nothing but a few Land Rover footprints on faraway Martian soil, and OK, OK, maybe the Land Rover, too. Astronomers are pure of heart and appealingly puerile. They look into the midnight sky and ask big questions, just as we did when we were in college: Who are we? Where do we come from? And why are we standing around outside on the night before finals, do we want to end up making elevator parts for a living like our father or what? Astronomers no longer have to worry about finals, although they do have to worry about getting their grants financed and their new telescopes built or at least not have the budgetary plug pulled on their old telescopes. In any event they are professional philosopeepers, and they ask the big questions about where we come from and what we are, and, much to their amazement, they have found answers to those questions swinging from the stars. Of the many extraordinary findings in astronomy over the past half century or so, space scientists cite two as cosmic standouts: the discovery and elucidation of the Big Bang that rang in our universe and the surprising centrality of ancient stars to the rise of life on Earth.

We may associate astronomy with the night and darkness, but one of the core truths of the discipline is its near complete dependence on light. "For astronomers, the universe is our laboratory, and the way we analyze what's going on in that laboratory is by analyzing light," said William Blair, a professor of astronomy at Johns Hopkins. "With rare exceptions, like the occasional asteroid or meteorite, we can't get our hands on the stuff we're studying, but we can learn a tremendous amount about the objects that are out there by examining the different types of light waves those objects emit, across the electromagnetic spectrum. This is one of the little details about the field that I don't think most people are aware of: that almost everything we've come to understand about the universe we have learned by studying light." Recall that optical light, the light waves that fall in the so-called visible range of the electromagnetic spectrum — the light that blinks back at us when we gape at the sky — represents a tiny segment of the light waves that astronomers study. They have designed a battery of bionic eyes capable of detecting virtually every radiant signal the firmament has to offer, from ultraviolet light through X-rays and out to the histrionics of high-energy gamma rays on the short end of the scale; and from infrared radiation, down into the misleadingly named microwaves, and over to the really lo-o-ong energy humps of radio waves. If you happened to see the 1997 movie *Contact,* in which Jodie Foster played a brave young astronomer battling genuinely astronomical odds and a generically asinine bureaucracy in her search for signs of extraterrestrial civilizations, you would have caught a few glimpses of the legendary Arecibo radio telescope built right into the mountainside of Puerto Rico. It's a huge telescope, 305 meters, or roughly 1,000 feet, across, and the diameter of the dish is a measure of just how wide are the radio waves it is set to sample.

By surveying the skies with instruments tuned to every possible wavelength of light, astronomers get a sense of what sort of cosmic bestiary we live in. Infrared telescopes can peer through the thick dust clouds that serve as a galaxy's stellar nursery and detect signals from embryonic stars within. Ultraviolet studies illuminate the nature of hot young massive stars, cool old dwarf stars, active galaxies, and hyperactive quasars. With X-ray and gamma-ray scans, scientists have probed black holes, pulsars, supernovas, and the mysterious gamma-ray bursters, thought to be an unusually violent class of exploding stars. Radio waves murmur hoarsely of the Big Bang from which all else sprang.

Beyond revealing its roots, each light beam speaks of the journey it had en route to its telescopic rendezvous: the relative desolation, dustiness, violence, or sedateness of the terrain it traversed, the masses it passed, the time it has been in transit, the likely fate of the radiant body that gave birth to it so very long ago. Another extraordinary ordinary truth of astronomy is that a look outward into space is also a look backward in time. Light is mighty light on its feet, and nothing in the universe is known to outrun it; but light is not infinitely fast, which means it needs time to get from point *a* to point *b*. And because space is so expansive, and the gaps between any two items so chasmic, the light we detect from the stars is old news. Even the light leaping off the surface of our nearest star, the sun, needs eight minutes to stream across 93 million vacuum-packed miles before it can strike your sensibly sunscreened skin. The image of Jupiter you see in your backyard telescope is how the planet looked half an hour ago, while that of Saturn is some seventy minutes old. Peer beyond our solar system, and you start digging into the lumen archives. In the constellation of Canis Major, for example, you'll find Sirius, the Dog Star, gleaming twice as brightly as any other star in the sky; those sequins left home nearly nine years ago. Or skip to the Little Dipper and mark the showpiece spot at the handle's tip; that light belongs to Polaris, the North Star, as it was back when Will Shakespeare was still wearing shorts.

Granted, nearly everything that you can perceive in the nighttime sky with the unaided eye very likely hasn't changed much between the time the radiant energy was dispensed and the time it reached Earth. On a really excellent night of stargazing, you can distinguish maybe 2,500 stars from any single location, and all of those stars, nearly all of the dots that the ancients connected on their charts into the named constellations, are located in our galaxy, most of them quite close, within a few hundred light-years of the sun. If it is dark enough and the right time of year, you can also see the fuzzy band of light that is colloquially referred to as the Milky Way, as though it had nothing to do with us, or the major and minor Dogs, or the double Dippers, or any of the other standbys of the night sky. Of course, once again you are navel-gazing, looking right at our home galaxy, this time toward the bulging central disk where most of the Milky Way's 300 billion stars reside. Thick plumes of interstellar gas and dust lying between the sun and the bulge obscure the view, but even if you could stare straight into the galactic heart, you wouldn't be looking terribly far: from our earthly perch, located about two-thirds of the way down one of the four major spiral arms of our pinwheeling Milky Way, it's only a distance of 26,000 light-years to the hub. There are a couple of other galaxies right at the edge of the bare eyeball's ability, most notably the Andromeda galaxy, located just to the south of Cassiopeia. Andromeda is much farther away than any of the visible stars, but still, it is the nearest big neighbor to the Milky Way, a mere 2.5 million light-years away. On a cosmic scale, where the average star manages to radiate more or less stably for several billion years, 2.5 million years is but a twink of an eye. So, yes, the star light you see tonight may be hundreds, thousands, a million years old, but with few exceptions the stars themselves are still out there, burning bright.

Spend some time with a serious telescope, however, and all bets are off, and so, too, are many of the lights. The stronger the telescope, the more distant the objects that astronomers can view. They can see far beyond the Milky Way, Andromeda, and the other members of our so-called Local Group of galaxies, to millions of other galaxies, tens of millions, hundreds of millions, billions of light-years away. They can see flocks and flocks of spiral galaxies shaped much like our own, whirls of cream aswirl in the black coffee of space; and elliptical galaxies that look like giant scoops of rice, with stars for grains; and riffs on the basic ellipse and spiral themes, along with rumpled deviants called irregulars — galaxies shaped like cartwheels, beer barrels, pork chops, and pencils, or those flat plastic monkeys that you hook into chains. Astronomers can also peer inside those distant galaxies and descry and tally their parts — their stars, their nebulae of dust and gas, even some evidence of planets and comets. They have found elfin galaxies of 100,000 stars, and colossi of 3 trillion. Whatever their shape and census, galaxies display an unmistakable coherence, their components clearly bound together by gravity into distinct communities, shining states of shared stellar fates. The word "galaxy" *means* "Milky Way," and fittingly so, for every one of the 100 billion known galaxies is, like ours, a place that stars call home — or used to. Remember that the more distant the galaxy glimpsed, the more archaic the image, and the more mind-teasing the implications. If you go to a good science museum or planetarium, you'll likely find some of the gorgeous "ultradeep field" surveys captured by the Hubble Space Telescope, of hundreds of extremely distant galaxies. With few exceptions, the stars in those galaxies pictured have long since died — sputtered and collapsed into dull brown dwarves, or spattered their outer sheaths into their surroundings, as supernovas. In some cases, hot new stars have taken the place of the ancestral lights captured by our telescopes. In others, the galaxies are likely cooler and darker and more sedate than they appear to our necessarily time-lagged eyes. Some of the galaxies are thought to have been swallowed up by surrounding galaxies, or by a giant black hole that lurks at their core, as black holes are thought to exist in the center of many galaxies, including our own.

In so many ways, deep-space scans can outspook a séance. For example, astronomers perpetually monitor the skies in search of supernovas and the abundance of data that such big light shows can offer. On average, a star explodes somewhere in a galaxy about once a century. To find these rare events, astronomers take weekly pictures of the same 8,000 or so galaxies, over and over again, Tuesday in, Marsday out. "We look for what's different," said Alex Filippenko. "Usually there's nothing, but every so often, we find a new, exploding star. Last year, we found eighty-two." One week, it's the same old barred spiral, with all the pizzazz of a Boise potato. The next, a blinding bombshell shatters the calm, swamping the rest of the galaxy's photonic sum. Can anything seem more instantaneous, more here and now and in your face, than a mammoth sun that goes kablooey? Yet once again, time bides its time and abides by the lawful, awful limits of light. The cataclysmic event that "suddenly" appeared on an astronomer's scan occurred, oh, half a billion years ago, and the "new" exploding star has long since dispersed into the void, and who knows that it didn't, in dying, seed the birth of another sun, with satellite Saturns and Jupiters and gazing Gaias of its own. On a cosmic scale, at least, there is always new hope from the dead.

The universe that we live in and are inextricably of was born nearly 14 billion years ago — 13.7 billion years ago, to be a bit more exact about it, and scientists are confident that this figure fits well with a welter of findings. The universe and everything it enfolds — all known and suspected matter and energy, all space and time, all broken dreams, lost loves, and inside-out umbrellas — began in the momentous moment we call the Big Bang. If the name sounds both a little smutty and a little

Barney, it should. When the great Sir Fred Hoyle coined the term during a radio interview some sixty years ago, he meant it as a glib putdown. An adamant atheist as well as a prominent cosmologist, Sir Fred disliked the idea then gaining currency of a universe with a defined origin, viewing it as the equivalent of a cosmic nativity scene open to any number of religious tie-ins; he and his like-minded peers favored a "steady-state" model of a static universe that had always existed in pretty much its current contours. Hoyle's heckle proved so catchy, however, that soon proponents as well as critics were referring to the hypothesized birth of the universe as the Big Bang. Even as an ever accreting body of evidence has transformed a plausible conjecture into a bedrock premise of contemporary space science, still the lighthearted tag line holds. True, there wasn't really a bang. A bang is a sound, and sound waves need air molecules to propagate, and in the beginning not only was there no air there, there were no molecules, either, or atoms, just pure energy.

And "big"? In the beginning, there was really the smallest small of all, the entire universe contained in something less than a billionth of a trillionth the size of an atomic nucleus. But let's be serious. An event like the birth of the universe is a very big deal, and it was a bang in the sense of being an explosion. A tremendous amount of stuff, of energy, the beginnings of matter and, importantly, of space itself, of somethingness rather than the unnerving utter nothingness that might have been, or not have been, broke free of its confinement, of an infinitesimally small and circumscribed borderline called a singularity, and began ballooning outward in all directions with unthinkable force and at relativistic speed — that is, close to the speed of light. So, yes, it *was* a Big Bang, and we can be glad that Hoyle chose to poke fun at the concept long enough to capture its dispositional humor in a phrase.

We don't know why there was a Big Bang — what preceded it, what triggered it, or what was going on at the moment of truth, moment oh-point . . . whoa. Using mathematical models, scientists have chased the universe back to a point hootingly close to the Big Bang — "to 10^{-35} seconds after time zero," according to Alan Guth, a physicist at MIT. But filling in that last little tittle, that niggling hundred-billionth of a yoctosecond, now that's tough. To resolve that, scientists will have to settle some difficult questions, like whether the laws of physics were born with the Big Bang and therefore collapse into meaninglessness when you metaphorically enter the singularity of the Big Bang; or whether the laws predated the Bang and perhaps gave rise to it. Whatever the cause, we know the consequences. Our universe began with the

Big Bang, and it has been expanding, and cooling, ever since; and everything about the structure, shape, and makeup of the cosmos — its silky homogeneity on a large scale, its lumpy clotting into stars and galaxies when you take a closer look — traces back to that moment of infinitesimally huge and splendidly, blessedly adulterated unity. Simon Singh, a physicist and science writer, has described the detection of the Big Bang as "the most important discovery of all time," and he may be right. But whereas other important advances like sliced bread and Teflon prove their case with French toast, what are we to make of this contender for science's ultimate laurel? We can't touch it, taste it, see it, or butter it. Why should we believe that the Big Bang is true?

The formulation of the Big Bang model of the universe was an exercise in reverse psychology. First, astronomers realized that the universe was expanding outward in all directions, like an inflating balloon, or a loaf of yeasty bread baking, or one of those Japanese paper flowers that unfurls in water; then they began working backward. Run a movie of those everyday analogies in reverse, and what would you see? A buoyant, globular party favor collapsing into a flat cat-tongue of rubber, or a hydrangea-sized blossom being sucked back into a slight, suspect pill. So, too, it seemed, would a backward film of the universe bring the scattered cosmic characters ever closer, until everything condensed into a very small speck of starter dough — if not quite as small as a single point, then at least a single point of view.

The man generally credited with discovering the expansion of the universe was Edwin P. Hubble, a legendary Missouri-born, pipe-smoking astronomer considered as comely as he was brilliant. He was "an Olympian," his wife declared, "tall, strong, and beautiful, with the shoulders of Hermes of Praxiteles." Hubble also managed the impressive trick of luxuriating in his considerable celebrity and hobnobbing with such nonastronomical luminaries as Douglas Fairbanks, Cole Porter, and Igor Stravinsky, all the while retaining his lofty scientific reputation. Fifty years after his death, Hubble's crossover appeal lives on, for not only do astronomers still apply "Hubble's law" and the "Hubble constant" to their workaday investigations, but NASA's decision to christen its multibillion-dollar space-based telescope the Hubble has helped to keep his name alive in the wider public eye, at least until the slowly decaying instrument itself winks off for good.

Hubble first won renown for his persuasive demonstration that our galaxy was not the be-all and after-all of the universe, and that many of the mysterious splotches on astronomers' photographic plates that had been dubbed nebulae, for their cloudlike appearance, were not constituents of the Milky Way, as the mainstream view had it, but were independent celestial bodies lying at staggering distances from our galaxy — bodies that were soon determined to be whole other galaxies. On characterizing these autonomous, shimmering star prefectures in greater detail, Hubble found evidence that not only were they really far away, but they were getting farther away all the time. Whichever galaxies in whatever quadrant of the cosmic landscape he examined, they all appeared to be fleeing from our poor little galaxy, as though the Milky Way had broken out in buboes, or asked for help with the dishes. Moreover, the farther the galaxy, the faster it seemed to be beating its retreat. This could be seen because in sprinting away each galaxy turned a bit red in the face, and the greater the distance, the deeper the beet.

Here we have one of astronomy's fundamentals, an essential food group of the field and a powerful piece of evidence in favor of the Big Bang model of the birth and evolution of the universe: the light waves coming from galaxies undergo what is called a redshift before presenting themselves at our door. If you compare telltale atomic fingerprints, or spectra, of the light from a distant galaxy with equivalent spectra of known light sources here on Earth, you'll see that the patterns of dark and bright lines on both sets of spectra are identical bar for bar, indicating that the same mix of atomic elements must be generating the radiation both out there and down here. On the galactic spectra, however, the entire array of lines looks as though it had been shoved over toward the redder, longer-wavelength end of the electromagnetic spectrum and away from the bluer, shorter-wavelength side, compared to the terrestrial benchmark lights. What does this redshifting mean? It means that the pulsating waves of star light, as they traverse the gulf between their natal galaxies and our vigilant scopes, are being stretched and pulled and lengthened, the distances between the crest and trough of each light wave gradually widened, the peaks softened, the pique assuaged.

To get a grip on why redshifting occurs, it might help to briefly train your sights on the sound of one train passing. As vigorously as the United States Congress has sought to eviscerate America's passenger rail system, you undoubtedly have had the pleasure of listening to the prolonged, poignant wail of a train whistle. If so, you surely noticed that the sound changed pitch as it rushed by. On approaching, the whistle had the high, grating pitch of a piccolo. At the moment of alignment between you and the whistling engine car, the sound dropped down to the midlevel root-a-toot-hoot of a conventional train whistle. And as the train left you in its rust-flecked dust, the whistle pitch dipped even deeper, finally assuming the mournful lowing of so long, take care, good-night to noises . . . everywhere.

To a pair of station-bound ears, the whistle's sonic profile changes so dramatically from processional to recessional that it's easy to forget how the whistle pitch sounds if you're on the train rather than missing it: pretty much the same from beginning to end. Only to a target in relative motion compared to the source of the sound does the famed Doppler effect come into play. The effect, named for the nineteenth-century Austrian mathematician and physicist who formulated it, says the waves generated by a moving object will shift in size depending on whether the object is headed toward or away from you. If the object is a noisemaker, the sound waves will compress into a higher pitch on approach and relax into a lower pitch on retreat. If the object is a floating leaf, the ripples spreading on water will appear more closely spaced if the leaf is drifting toward you than away from you. The redshifting of light detected in the study of distant galaxies, then, is just another instance of the Doppler effect at work.

Importantly, however, the light from distant galaxies is always displaced in one direction. In our immediate Local Cluster of galaxies, there is some give and take. The Andromeda galaxy, for example, is blueshifted, a persuasive sign that it is headed toward us and we toward it, and that in about 6 billion years the two galaxies will, as a result of our mutual gravitational attraction, merge into one. But step outside the neighborhood, and you'll see only red. As revealed by the unwavering augmentation of their light waves, the remote galaxies must all be racing away from us. What's more, the greater the distance of the galaxy, the more extreme its redshifting, and in a roughly proportional manner. That is, if Galaxy Wibbleton lies at twice the distance from Earth as Galaxy Wobbleton, the light from Wibbleton will be twice as redshifted, twice as elongated, as the light from Wobbleton; if three times as far, it will look three times as red. How to explain this link between galactic distance and redshift radicalism? According to Doppler's equations, the velocity of any wave-making body affects the relative spreading or narrowing of its waves. The whistle of a fast train will, to an outside listener, squeal higher as it approaches and moan lower as it recedes than that of a slow train. In fact, the correlation between velocity and degree of Doppler shiftiness is precisely what allows a police officer to gauge your driving speed by bouncing radar signals off your vehicle and seeing how severely the car's speed distorts the wavelengths of the incident radiation; the greater the Doppler shift observed, the bigger the ticket you'll be served. Those distant galaxies, in other words, must be retreating from us at a greater velocity than are the closer galaxies.

Or maybe we're just paranoid. As it turns out, the sense of being uniquely repulsive to almost everything in the universe is an illusion. If we were located in the Andromeda or Sombrero or M63 galaxy, the redshift profile of the cosmos would look the same as it does here in the Milky Way: as though other galaxies are all moving away from us, and at a velocity more or less proportional to their distance. How can this be, that each of us is cast in the role of mortician at a bridal shower? To understand the phenomenon, try this simple experiment, which requires nothing more than a balloon, a felt-tipped pen, and a second pair of lips. First, decorate the flaccid balloon with polka dots, spacing them out as evenly as you can. Then, ask your assistant lips to take up the newly freckled balloon and slowly blow. Put your finger on any of the dots, and study the dots that surround it. Note that as the balloon expands, the neighboring dots are all moving away from your finger. Note as well that the dots nearest your finger are retreating from you at a slower pace than are those farther away. The reason is that there is less expanding rubber between you and a neighboring dot than you and a far dot, less relative surface area to tug adjacent points from your vicinity. Now place your digit on one of those far-off dots and again look at the points around it. Same thing. The inflation of the landscape is pushing all spots out and away, out and away, and the far dots are receding from your finger more briskly than those close at hand. OK, that's enough blowing for now, Grandma, can I get you a cup of tea, seltzer, oxygen tent?

The expanding universe is not that different from an expanding balloon, except that the universe is bigger, colder, and darker, and it won't pop even if you put it in a cage with a pair of mating ferrets. Still, the balloon analogy shows how every vantage point on its swelling terrain appears to be the center of the universe without really being the center of the universe, and how distant objects will recede from one's vantage point at a faster velocity than near points not because they're sailing faster in any "real" or absolute sense but merely relative to the recession rate of sites nearby. The galaxies at greatest distance from Earth are not Olympian sprinters, not Hermes of Praxiteles to the local Yertles of Sala-ma-Sond. Their spectacular velocity is spectacular only to us, while to one another they are cruising along at unremarkable speeds. As Albert Einstein demonstrated in his special theory of relativity, it is meaningless to speak of an object's absolute speed or motion through space, for there is no final arbiter, no unchanging, eternal grid against which that speed can be clocked. All you can ask is "fast compared to what?" From our perspective, we and our nearby galaxies are moving through space at about 370 miles, or 590 kilometers, per second, which is only slightly faster than a tractor-trailer headed down a Montana freeway at 2:00 in the morning. By contrast, the most distant galaxies seem to be receding from us at velocities of thousands or tens of thousands of miles per second, uncomfortably close to the speed of light and illegal even on the Autobahn. To their local highway patrolmen, however, those far-off galaxies are tooling along at disappointingly licit speeds of roughly 590 kilometers per second.

Another way in which the balloon exercise can illuminate the nature of our expanding universe is this: the dots are not really taking the initiative and moving away from one another, as they might if they were ants on the surface rather than pen markings embedded in the surface. Rather, the skin between the dots is stretching wider. Similarly, the galaxies of our universe are not really rushing away from one another. They're not traveling through space, they're traveling with space. They are more or less staying put, while the space between them just keeps expanding. This distinguishes large-scale galactic motions from other celestial pilgrimages. Under the spur of gravity, the Earth and its sibling planets orbit the sun. Our solar system in turn is gradually making its way around the dense and gravitationally convincing hub of the Milky Way, completing a galactic circumgyration once every 230 million years. But while there are some regional exceptions (like the gravitational attraction that is slowly drawing us and Andromeda closer together), galaxies are distributed through the cosmos with sufficient homogeneity that they end up being gravitationally neutral with respect to one another. The galaxies themselves are neither wandering nor widening. It is the space in between that can't stop loosening its belt.

I fear that, on a visceral level, this idea is almost impossible to accept, no matter how many packages of balloons your loyal, rapidly blue-shifting Nana inflates — that the expansion of the universe is not a tangible matter of galaxies exploding outward into space like shrapnel from a bomb, but of space itself exploding outward, the shrapnel trapped in its hide. For one thing, space is not supposed to do anything except sit there waiting to be crossed or filled. For another, what is it expanding *into?* More space? If so, why doesn't all the space just smear together in the first place? How could we have an expanding universe if space is expanding into space? Wouldn't that be like trying to blow up a balloon full of holes? Well, you may derive some comfort in knowing that astronomers don't have an intuitive grasp of the subject either.

"The expansion of space is a concept that I understand mathematically, but on a personal level, no, I can't do it," said Mario Mateo, a professor of astronomy at the University of Michigan.

In fact, when Raman Sundrum, a professor of physics and astronomy at Johns Hopkins, gives public talks about universal expansion "and where all the galaxies are going," he has his audience do a modest mathematical exercise. Think about simple, whole numbers, he says — one, two, three, four, five, and so on. How far apart are these numbers from one another? One, his listeners reply. The distance between them is one. And how many such numbers are there? Well, they keep going, the audience responds. There's an infinite number. "Now I tell people to double every number, so that one becomes two, two becomes four, three becomes six, and on and on," said Sundrum. "For every original number, I've given you a new number, but the distance between them has gotten bigger. Now I tell you to double the numbers again, to four, eight, twelve, sixteen, twenty. And again, to eight, sixteen, twenty-four, thirty-two. I'm always giving you a new number for the old number, so that it's the same number of numbers, but the distance between the numbers is getting further and further apart. In a sense, the numbers are receding from each other faster and faster, and what we're seeing with galaxies is something like that. As far as we know, the cosmos can extend forever, just like the numbers can go on forever. What are the cosmos expanding into? Well, what are our numbers expanding into? There's no running out of space, and there's no running up against an edge, and yet, the distance between the galaxies, as with our integers, keeps increasing."

Perhaps easier on the mind's eye than the unseemly uppitiness of expanding space is to do as cosmologists have and run the tape in reverse. If you systematically rein in all those receding galaxies, at speeds counterproportional to the velocities with which they are disappearing into the distance, you eventually reach a point where they're piled one on top of the other: the equivalent of some 100 billion galaxies of hundreds of billions of stars apiece, all cohabiting the same pre-place, ante-space patch of proto–real estate. The lights, and the plasma, and the inferno come free.

After discovering galactic redshifts and connecting and reconverging the dots backward toward an event that, for all the hubristic ring of it, could be thought of only as the birth date of the cosmos, scientists began sketching out what conditions must have been like when the universe was new. Their computational conjurings proved materially as well as aesthetically fruitful, for they eventually yielded the second major piece of evidence in support of the Big Bang, as we'll discuss in a moment. So what might our big fat bouncing pistol of a neonate have looked like? First of all, cosmologists plead, bear in mind that the birth of the universe didn't take place in a specific location in space because the space and matter of our universe came into being simultaneously, essentially bubbled up out of . . . well, we don't know what. The void? Another bubble in a larger burbling crockpot of cosmic chowder, of universes within universes? We don't know, and we may never know, for what is beyond our universe may remain forever inaccessible to any sensors or instruments within our universe; and without evidence, we are trading not in the science of astrophysics, but in idle metaphysics, sophomoric philosophistry, and a few too many boxes of Milk Duds.

In any event, what we do have evidence for is this: In the beginning, there was light all right, an overwhelmingly bright, hot light like nothing you've seen or felt or could see or feel, for, as Alan Guth so gaily put it, "The photosensors in your eyes would evaporate instantly." The light! the light! exploded into being, a radiant seed of pure energy, tinier than the proton of an atom yet almost infinitely dense, with a temperature in the trillions of degrees, and it instantaneously began swelling outward. Almost immediately after the expansion had begun, some of the energy managed to condense into matter, into elementary particles, like electrons and the subparts of protons and neutrons, the quarks; and their countercharged, counterspinning antimatter counterparts, the positrons and the antiquarks.

Still the mat of matter and energy expanded mightily. In a fraction of a trillionth of a second, the universe had inflated from its subatomic birth girth to something on par with a cantaloupe; before a thousandth of a second had passed, the cosmos had cast itself across two-thirds of a mile. It grew and it glowed with a light not only blindingly, deretinizingly bright, but of a purity and uniformity not seen in our quotidian lights, our lamps or our suns or our bombs. There were, in fact, tiny ripples in this dawn's early light, minute irregularities in the radiant paste of the universe that eventually would prove our salvation, but these gradations, these flickering slubs, were of quantum amplitudes, and so the light at first seemed unimpeachably pure, fittingly ethereal.

For the firstborn particles, the swelling belly of the baby was pure hellion, and they were smashed and rattled and ashed into radiation and reparticulated back out again, over and over. Nevertheless, with the expansion came a deruffling of tempers, a sufficient cooling for the condensation of matter to continue beyond the most elementary phase.

Quarks teamed up as triplets into reasonably sturdy protons and neutrons, while nearly equal numbers of triumvirates of antiquarks formed into antiprotons; and through the archaeostew flew electrons and positrons, too. Still the material surl was not through, for matter and antimatter cannot share the same province and survive. Protons and antiprotons sprang into being, only to collide into mutual annihilation; electrons and positrons were tossed together and lost together. Luckily, for reasons that remain mysterious, the early universe had been salted with a slight excess of matter over antimatter: for every billion or so antiprotons and positrons aswirl in the starter stew, there were a billion and one protons and electrons. The result? When the great matter-antimatter hissing match had finally played itself out, there were just enough protons and electrons left to start building atoms, stars, galaxies, cats, hats, pianos, piano tuners, physicists, and atom smashers to recapitulate conditions of the early universe.

Even with antimatter effectively neutered, the universe would need close to another half a million years before it was fit to be seen. Until then, all was a fog. The universe was still so hot and dense that matter could exist only as a plasma, a sea of nuclear particles and footloose electrons that scattered the light every which way, as do the water molecules of a fog or cloud. "A plasma is very nontransparent to electromagnetic radiation," explained Alan Guth. "In our early universe, photons of light were constantly colliding with the free electrons and bouncing off in different directions, so the radiation didn't get anywhere in this period." And just as it is practically impossible to peer into the heart of a thick cloud, so astronomers suspect that the plasmic conditions of the early universe rule out any hope of detecting the electromagnetic signals of the Big Bang proper.

After 300,000 years, however, the fog began to lift. The universe had expanded to a diameter about $1/_{1,500}$ its current size, and it dropped to a temperature of a mere 3000 degrees, cool enough for electrons and protons to begin expressing their innate compatibility, their electromagnetic complementarity, and to form electrically neutral atoms — simple atoms like hydrogen and helium, but full-fledged, thoroughly modern atoms nonetheless. The opacity of plasma finally gave way to the transparency of a gas. At last the universe's radiant energy could begin traveling outward in a straight line, rather than being restirred back into the pitiless plasmic paste, and it has been flying freely ever since.

Adherents of the Big Bang model of the universe proposed in the 1940s that it should be possible to detect the boundary between the opaque early universe and the transparent universe that has reigned ever since, just as we can look through a clear sky over to the edge of a giant cloud formation. This boundary is what they call the "surface of last scattering," or "the wall of light" — the last time in the history of the universe that matter managed to smear the astral radiance into a milky blur. The wall of light should be all around us, they said, because it is the relic glow of the whole universe as it was when it was much smaller, but that universe has since puffed out about us, as it has about all the other dots on the balloon, raisins in the cake, and the like. Or picture being aboard a raisin in the middle of a thick puff of smoke, which then expands outward from us like a spherical, shimmering smoke ring. Wider and wider it billows. Time passes, and now we're standing down here in the clear, contentedly fruit-bound, peering through a large volume of transparent space, looking outward for the halo of fuzz that once upon a time was all that was.

Yes, the fuzz has to be out there, Big Bang theorists suggested. It's part of the universe; where else could it go? They also calculated that the radiation billowing off this 3000-degree surface of last scattering would have started out as extremely high energy, which means of an extremely shortwave variety. In the intervening billions of years, however, the light, as it is wont to do with lengthy travels, would have redshifted down, down, down to long, cool wavelengths, in the long, ruddy end of the microwave portion of the electromagnetic spectrum — down to wavelengths you'd expect to be emitted by a radiant body not of 3,000 degrees, but of 3 degrees. In the mid-1960s, astronomers at Bell Labs in New Jersey detected the ambient blush, the remnants of the plasmic early universe, at the predicted 3-degree wavelength, an achievement for which they were awarded the Nobel Prize. The sense-a-round radiation is known formally as the cosmic microwave background, and you can detect it yourself, in the comfort of your home, especially if you don't have a decent cable connection: the snowy interference that appears on your untuned TV is partly the result of the cosmic microwave background, the cold crackling leftover light from the universe in circa 300,000 A.B.B. It is the first fossil, the earliest snapshot, and, if not quite music to one's ears, the closest thing we have to the music of the spheres. Together, the cosmic background radiation and the redshifting of distant galaxies sing softly but surely of a very Big Bang, and of an expansion that began 14 billion years back and just keeps Beguining along.

The cosmic microwave background is ubiquitous, and impressively uniform. The night sky might look different when you're in the outback of Australia from the way it does when you're in Halifax, Nova Scotia, with a different arrangement of constellations, but the cosmic microwave signal you'd pick up in either location would be almost identical in strength and wavelength. That sameness attests to how much more uniform in temperature the universe was in its smaller and more compact past than it is in its current state of middle-aged spread and diffusion, and logically so: Think of how much easier it is to heat a small room evenly than a large, drafty Victorian house. The uniformity of the microwave signal also underscores the uniformity with which the universe has expanded since the Bang, or at least in the eons since the plasma age ended. The smoke ring has been pushed out the same amount in all directions, and so we perceive the same sort of cool, radiant signal from all directions.

It turns out, however, that microwave radiation is not completely homogeneous. Probing the celestial lightscape with sensitive instruments borne aloft by satellite and — how appropriate — balloon, astronomers have found minute fluctuations in the background microwave radiation, spots where the signal is comparatively stronger or its wavelength longer. To cosmologists, such flickerings in the wall of light indicate that the mass of the early universe — the stuff that was bouncing the universe's radiation into a plasmic haze — was not perfectly, smoothly distributed. From the moment that matter materialized, they said, it had a little clumpiness to it, the result of so-called quantum fluctuations arising from the natural jitteriness of subatomic particles. In other words, cosmologists argue, the basal canvas had no choice but to ripple; the laws of physics, the probabilistic nature of quantum mechanics, demanded it. And those tiny ripples that we see today as trifling trembles in an otherwise standardized background glow were likely the source of all cosmic diversity and opportunity. "Those ripples were responsible for the formation of galaxies, stars, of universal structure generally," said Guth. "Without them, the universe would have been a giant cloud of hydrogen gas, and a very dull place indeed."

Our universe was certainly no idle gasbag. From the start, it had a kind of cytoskeletal integrity to it, filaments of comparative density that only gained in strength and intensity as the universe grew. Over the next few hundred million years, the first stars and galaxies began to condense out of those pockets of comparative density in the expanding plume of atoms and energy. And though galaxies today are the only known homes for stars, the only place where stars are born and die, and you won't find any hermit stars wandering through the gravitationally barren wilderness of intergalactic space, that doesn't mean the galaxies got there first. After all, who better to know how to build a warm home, a thriving community, a stellar society, than the residents themselves?

Astronomers have much to learn about the evolution of cosmic structure, but they now suspect that stars may well have preceded galaxies as the earliest celestial bodies to form out of the spidery gaseous mass of the young universe. Not just any stars, though. Not stars like our sun, much as we love it and are lucky to have it exactly as it is. Rather, the first stars were likely to have been massive, thousands of times bigger than ours. Giant stars alone have the power of which Isaac Newton dreamed before a falling apple so rudely, apocryphally awoke him — the power of alchemy. Giant stars alone can start with the simplest, lightest atoms, like hydrogen and helium, and forge them into the whole periodic palette of the elements, into all the Rubenesque beauties with their thickset nuclei — nickel, copper, zinc and krypton, silver, platinum and gold, and tungsten and tantalum and, yes, mercury and lead. We humans are not unique in our greed for all that glitters. Once seeded with traces of heavy metals, the gift of those founding stellar magi, the wider gaseous terrain of the early universe began taking on shape, and, in very short order, the skies were ablaze with millions of stars living in distinct Milky Ways.

Which brings us to another of the great discoveries of modern astronomy: Joni Mitchell was right about us all being stardust. Our lives depend now on a single, living sun, but other suns before this one have died to give us life.

The observable universe may be more than a shapeless cloud of hydrogen gas, but nevertheless this least frilly of all elements is by far the most common. Nearly three-quarters of ordinary matter consists of hydrogen, the atom with one proton and one electron to claim. Helium, the second character on Mendeleev's table and possessor of two protons and two neutrons at its core, accounts for about 24 percent of known matter. All the hydrogen and much of the helium in existence today, along with a sprinkling of the universal stock of lithium, boron, and beryllium atoms, are the direct product of the Big Bang, generated when the universe was new. The next time a family member brings home one of those unsightly Mylar balloons that remains stubbornly, desultorily afloat for so long that you decide to shred it when the balloon's owner isn't looking, consider that at least a few of the helium atoms you are about to blithely disperse into the atmosphere may have been around in their current configuration for 13.7 billion years. Now hurry up and get rid of that thing before the kid comes back.

Ambitious though the Big Bang was, its inventive capacity was limited and short-lived. The laws of physics, which either preceded the Great Explosion or were born with it, dictate that the electromagnetic force will keep the positively charged protons of discrete hydrogen nuclei as far as possible from one another, unless something pushes them so close together that the strong force can take over. As the mightiest known force in the universe, the strong force can strong-arm the innately xenophobic hydrogen nuclei until they agree to fuse into something new — into atoms of helium. Or it can fuse together helium and hydrogen atoms into an even bigger nuclear commonwealth, a state called "being lithium." Yet each ratcheting up in the size of the atomic polity requires that much more heat and density to manage, that much more ambient extremity to overcome electromagnetic repulsion and allow the strong force to work its diplomacy. The Big Bang got as far as impelling packets of five protons into meaningful proximity — into a scattering of beryllium atoms — before its mass had dispersed and its initial pressure-cooker conditions had dipped below those needed to drive fusion's fancies. In the wake of the natal stabs at nation-building, the bulk of the universe's atoms remained in the same parochial hydrogen format in which they'd begun.

Yet all atom-making was not through. There were those quantum-borne quivers, those clumps in the cloud, and there was gravity, gracious, warm-hearted gravity, with its sensible shoes and its feet on the ground. Gravity is the weakest of nature's four forces, but it works well on large masses, and it has the added advantage of always attracting, never driving away. After a million or so years of breathless expansion driven by the Bang's phenomenal outward-bound pressure, gravity began to exert a moderating counterforce. The pace of wholesale, whole-scale growth slowed ever so slightly, allowing the denser pockets of matter in the universe a chance to dawdle, churn, swirl, and circumnavigate. And once a sufficiently dense exemplar of these hydrogen redoubts started twisting around on itself, gravity got a really firm grip, and pulled the gaseous pocket inward, into a ball. As it condensed, the gas grew hotter, its atoms more agitated. Soon it grew so hot that the electrons were again stripped away from their nuclear partners, returning the gas to the plasmic state of the small, early universe. In the center of the gaseous orb, where the heat and pressure were at maximum exhortation, not only were the electrons torn free from their protons, but the protons of the individual hydrogen atoms were squeezed closer and closer together, until finally their mutual electromagnetic repulsion could be overcome and the business of nuclear fusion could begin anew — on a bolder and more ambitious scale than anything seen in the Bang's early pangs.

As we all know from years of hearing the energy industry pine for the power to tame it, nuclear fusion is a wondrous thing, a Rumpelstiltskin on stilts. Not only does it transform the light and the simple into the weighty and complex, but the very act of fusing atomic nuclei together releases a big jolt of electromagnetic radiation — of energy. We humans have succeeded in fusing together hydrogen atoms and unleashing a tremendous blast of energy in the process: that is the source of a hydrogen bomb's apocalyptic power. Far trickier is figuring out how to fuse atoms in a controlled, orderly, and, of course, cost-effective manner. It's a daunting task, but one that our sun and the billion trillion or so other stars in the universe accomplish every day. The source of a star's energy, its shine, its heat, its guiding wishworthy light, is thermonuclear fusion, the perpetual merging at the dense stellar core of large numbers of small atoms into a smaller number of larger atoms. The power of nuclear fusion is the defining hallmark of a star, and it takes a certain heft and density to pull it off. Jupiter is a very big ball of gas, but it's not quite big enough. The atoms at its core are not under sufficient pressure to change their elemental identity. Only at a mass about eighty times greater than Jupiter's will a ball of gas have the stoutness of heart to accomplish thermonuclear fusion, the squeezing together of reluctant singleton nuclei into the radiance of atomic matrimony.

The first stars to condense out of the primal nebula, though, were likely much bigger than eighty Jupiters, or even eight hundred suns, for as they began their collapse, the compaction through gravity of a thick slub of gas into a tidier and more coherent sphere, the increasingly dense object would attract ever more matter from its dusty surroundings and so grow huge rapidly by accretion. The early universe was a cramped, cluttered, dusty, gaseous place compared to today, and so a condensing ball had no choice but to pull in huge hanks of extra matter as it tightened in on itself — to augment its mass even as its volume diminished. That hugeness exacted a high personal cost: giant stars die young and violently. Yet if their lives were brief, their art would long survive them, and it is well worth a look at the docudrama of stellar genius.

For simplicity's sake, let us assume that our model founder star is made of pure hydrogen, with none of the other elemental efforts of the Big Bang adulterating it. Our exemplar star is an enormous condensation of hydrogen, hundreds of times more massive than the sun, and the electrons have been stripped from their protons, and all is a plasmic bisque. Gravity is tugging everything inward, toward an imaginary point at the center, and so the pileup of hydrogen particles is greater the more deeply you delve. At the superhot, high-pressure confines of the core, the hydrogen nuclei are swirled and squeezed, swirled and squeezed, until a critical threshold is surmounted, electromagnetic repulsion is defeated, and discrete hydrogen particles are fused into helium nuclei. The energy liberated by this thermonuclear fusion begins radiating outward, from the core toward the surface, and the billowing up of heat and light offers a counterbalance to the inward pull of gravity. In fact, the pulsing radiation, the bounty of fusion, is what keeps the star intact, keeps its inner layers from collapsing under the weight of those on top. But the effort is energy-intensive and cannibalistic.

As the Oxford chemist Peter Atkins has written, a star's appetite for hydrogen is "truly prodigious." Our sun, for example, fuses 700 million tons of hydrogen into helium every second, and in so doing radiates away pieces of itself each day, splashing warmth and light across the solar system, and My Very Educated Mother and her Nine (or eight) Pies, and their retinue of moons, and the asteroid belt, and Hale-Bopp, and Comet Kohoutek, too. Yet even though the sun has burned for 5 billion years, and even as it grows gaunter with every passing moment, it has enough hydrogen, packaged with just the right degree of density, to stay lit for another 5 billion years.

No such longevity for our voluptuous stellar forebear. Here, the great weight of its mountains of accreted matter heats up the core layers with staggering rapidity, accelerating the pace of fusion and quickly depleting the star's hydrogen stores, perhaps in as little as a couple of million years after the star first formed. Its hydrogen fuel exhausted and the stabilizing counterpressure of fusion energy momentarily cut off, the star again falls prey to gravity and sharply contracts. That downsizing in turn raises the temperature and density of the core, until the next thermonuclear threshold is crossed. Now the helium particles, the fruit of the previous act of fusion cuisine, start merging into carbon, infusing the star with a fresh burst of radiant energy that thwarts gravitational collapse. That is, until the helium, too, is spent, at which point another round of contraction ensues, followed by a new fusion event, and the creation, the so-called nucleosynthesis, of still heavier atoms at the stellar core. Onward through the periodic table of the elements the star marches in its struggle to stave off total collapse, hammering smaller nuclei into nitrogen, oxygen, sodium, phosphorus, potassium, calcium, silicon, yes, the familiar ingredients listed on nutrition labels or kicked up by bullies on the beach; onward through to the nucleosynthesis of iron and nickel, the elements with the most stable nuclear configurations of all. We're only about a quarter of the way through the atomic table, and there are many other, heavier elements yet to be synthesized, but iron and nickel mark the end of the line for fusion power. If you fuse an iron nucleus with another nucleus, you won't release any energy. To the contrary, this heavy-handed union requires an *input* of energy. And the fluttering outward of liberated, radiant energy is what keeps a star from caving in on itself. The last good-bye is nigh.

At this stage, the star is a like a great ball of baklava, with a dense center of nickel and iron nuclei surrounded by thin shells of successively lighter elements that it had baked up through the ages but hadn't gotten around to cannibalizing. Lacking any radiant bulwark against gravity, the whole construct again condenses, and the core temperature soars to 8 billion degrees, hot enough to synthesize elements a bit beyond iron and nickel, yes, but to no avail for the star: its engine of thermonuclear stability, the release of radiant energy through fusion, is dead. The core begins to lose structure, and upper layers dive in toward lower layers. Violent photons of light ricochet in all directions, splitting apart any heavy nuclear particles that stand in their way. The star's interior goes into free fall, the stippled, plasmic strata of itself streaming helplessly toward that imaginary point at the center of the orb. In less than a second, a core the width of many suns is squeezed down to something the size of North America. The catastrophic contraction sends shock waves through the entire celestial body and blows out a halo of stellar matter "like a great spherical tsunami," as Peter Atkins puts it. Our star explodes as a supernova, and in those furious closing moments of its life, the real heavyweights of the elemental table are forged — platinum, thallium, bismuth, lead, tungsten, gold. The newborn particles are scattered into space, along with the many other, comparatively lighter elements that the star belly had built up before the whole star went belly-up.

For this sequined shrapnel, we can thank our lucky stars. By salting the young universe with heavy elements, particularly metals, the first few meganovas helped touch off a boom in stellar construction. The background gas was hot, Chuck Steidel explained, and it's difficult to get stars started from masses of overheated, overexcited gas. The metal particles bequeathed by the progenitor stars cooled the gaseous landscape enough for a multitude of nebulous eddies to begin condensing into stars, into clusters of stars, into bustling, burgeoning barrios of stars. "We think it went from massive stars to small galaxies relatively quickly, during the first billion or so years of the universe," said Steidel. Larger galaxies would then have formed through mergers and acquisitions, by collisions between galaxies or by one comparatively dense galaxy gravitationally sucking in the contents of a smaller galaxy. By 12 billion or so years ago — 1.7 billion years after the Big Bang — the majority of the universe's galaxies had formed, including our Milky Way, although they would continue to sail ever outward and away from one another, buoyed on the expanding silk of space; and each galaxy would continue to evolve, its gathered goods to rotate around the midpoint of its mass, its stellar citizens to live out their lives at varying tempos and temperatures, depending on their mass and their proximity to other stars. In many galaxies, particularly spiral ones, we find flourishing stellar nurseries, comparatively thick patches of gas and dust from which new stars perpetually are condensing, the natal event often driven by the obliging and violent death of a massive older star that lived next door.

So it likely was with our solar system. Some 5 billion years ago, the shock waves of an exploding supernova and the concomitant expulsion of the star's salubrious heavy elements into interstellar space spurred a ragged cloud of gas and dust in one of the Milky Way's arms to begin condensing. As it contracted, the nebula began to spin (just as a figure skater spins by pulling in her arms) and to flatten into a disk (as a figure skater happily does not). Through several million gyrating years, the bulk of the mass was drawn by gravity toward the center of the pancake, forming a bulge of ever escalating heat and density, which finally burst into thermonuclear splendor. Still some discus matter remained around our newborn sun, a petticoat of gas, dust, and all the hundred-odd elements that Dmitri Mendeleev would later seat around his table. That matter formed clumps: the protoplanets and their protomoons. Closer in to the central orb, only aggregates of rock and metal could withstand the heat, and so the four inner planets — Mercury, Venus, Earth, and Mars — are balls of rock and metal and are designated the terrestrial planets, firma to their core. Farther out on the disk, it was frigid enough for water to freeze, and once the ice particles had formed they collided and gathered gas and dust in a veritable snowball effect, yielding the four outer planets, the so-called gas giants — Jupiter, Saturn, Uranus, and Neptune. As for Pluto and Sedna and others of their subcompact class, whether you consider them planets, dwarf planets, planetisimals, planet parodies, or Planters party mix, they were formed in the Kuiper belt, one of the coldest, flimsiest nethermost rims of the solar disk, where there was not enough there to make much of them. Pluto and Sedna are among the behemoths of the icy, rocky bodies in the belt, and still you could hide nearly 10 Plutos inside tiny Mercury, and maybe 150 inside the Earth.

Our sun is a good star, a stalwart star, and its life span is only halfway through. But when its hydrogen store starts running low, the sun will have just a few ploys for keeping its plasma aflame. In 5 billion years, having depleted the hydrogen at its dense core, the sun will start burning the hydrogen in its comparatively thin outer layers, puffing up as it does so to thirty times its current girth. That swollen sun will be a cooler sun, its radiance ruddier than today. Our sun will be a red giant, and woe to any earthlings who may be around to witness its bloated blush, for the planet on which they stand will likely be vaporized in the expansion. Our distant descendants would do best to abandon Earth well ahead of time, relocating to, say, one of the bigger moons of Jupiter or Saturn. Come the sun's expansion, places like Jupiter's Ganymede and the saturnine Titan will be transformed into far more clement places than they are today — their skies brightened, their ice stores thawed into liquid oceans and rivers. Titan even has a gas atmosphere that, while not currently breathable, in theory could be reconfigured to suit human respiration, and its scenic views of Saturn's rings are an obvious plus. Wherever the space farers alight, they may as well kick off their boots and settle into a comfortable chair. The sun will radiate as a red giant for another 2 billion years.

And then? Then it's time to decamp and head for a whole new solar system. Our star lacks sufficient mass to explode, and instead it will simply sputter into barren obscurity. After the hydrogen shell has been exhausted, the core will contract sharply, and its upper layers will start sloughing off into space. In the end, all that will remain is a dense, smoldering ember of carbon and oxygen barely bigger than Earth. The once mighty Ra and plucky *pro tem* red giant will have become a white dwarf; and though it can no longer generate fusion power, by sheer heat it still glows, and it will glow in this guise for the rest of all time.

The sun and other midsize stars can build from the Bang basics a handful of the elements we bio-Legos demand, notably carbon, oxygen, and nitrogen. The frequency with which ordinary stars concoct oxygen partly explains why oxygen is the third commonest element in the universe, after hydrogen and helium; and the combined commonness of hydrogen and oxygen explains why there is water, water everywhere, though only on Earth in abundant drops to drink. But stars of modest means and restrained temperaments keep most of what they make to themselves and supply only trace quantities to the universal inventory of post-helium parts, the weighty elements from which animate matter is made. The overwhelming bulk of our mortal cargo — the carbon in our cells, the calcium in our bones, the iron in our blood, the electrolytes of sodium and potassium that allow our hearts to beat and our brain cells to fire — was stoked in the furnaces of far larger stars than ours and splattered into the cosmic compost when those stars exploded. "We are star stuff, a part of the cosmos," said Alex Filippenko. "I'm not just speaking generically or metaphorically here. The specific atoms in every cell of your body, my body, my son's body, the body of your pet cat, were cooked up inside massive stars. To me, that is one of the most amazing conclusions in the history of science, and I want everybody to know about it."

The gaseous nebula from which our solar system formed very likely had been enriched several times over with star stuff, with the luxurious carnage of multiple supernovas that had exploded nearby over the course of the last 10 billion years. Each round of enrichment had enhanced the chance that the cloud at last would cool, and swirl, and condense into a skirted star, and the skirt would prove elementally weighty enough to yield the rocky, complex inner planets on which life could make a deal. Not on eenie or meenie, and Moe, I don't think so. But behind curtain number three, the full monty, it's mine.

We know that there is life on Earth, and that at least one species among its phylogenetic plentitude is, if not always sensible or reliable, certainly very clever at inventing tools, especially tools that allow us to engage in animated, disembodied forms of communication while simultaneously driving, jaywalking, or attending our daughter's piano recital. We are such indefatigable telecommunicators that the world and its 6.5 billion content providers don't feel like enough, and we can't help but wonder, Who else can we call? Are there other beings, on other worlds, and will we ever be able to contact them, or they us? Are we alone, or one of millions of habited planets in the galaxy, or billions in the universe? Will it ever stop feeling so hard and so hollow to ask and to ask and to ask again? Is there any evidence one way or the other for extraterrestrial life? What do astronomers think, and does their thinking on this most cosmic of all questions have any more moment than a five-year-old's musical Milky Way dream?

The answers to these questions are a mix of bad news, no news, and good news. The bad news is, no, we can't yet contact any extraterrestrial beings, not even with the sort of miraculous long-distance connections by which presidents ring up astronauts to banter about space food and mountaineers trapped in a storm on the top of Everest call their loved ones to discuss the slim odds of their being home for dinner. If we could, don't you think they'd already be working as customer service representatives under suspiciously bland names like Hank or Sherry?

Most of the news from the space-alien front is, alas, no news, or rather we-don't-know news. We have no evidence one way or another about whether there is life on other worlds. None. After the initial flurry of excitement in the 1990s over the possibility that we had detected signs of past or current microbial life on Mars, the evidence fell apart. There is no credible evidence that extraterrestrials have ever visited Earth or abducted any earthlings or searched any earthly body cavities for their inscrutable, nefarious purposes. Aliens have yet to respond to the recordings we included aboard the first two *Voyager* spacecraft, launched in 1977 — of wistful greetings in fifty-five languages; the music of Bach, Beethoven, Louis Armstrong, Peruvian pan pipers, Azerbaijani balaban players; and whales singing, chimpanzees grunting, and a train whistle passing in exemplary Doppler form. Is there life on other planets? We don't know yay or nay, there's no proof either way, so on this topic scientists can have nothing to say — can they?

No they can't. And yet, they do. The good news, such as it is, and I warn you it isn't much, is that the great majority of astronomers I interviewed believe there is life on other planets. Some think that life is common, that the universe is flooded with star stuff stuffed into self-replicating organisms of more or less cellularly based structures. Others say that life is likely to be rare, but that nonetheless it's probably not limited to Earth. Their conviction comes down to sheer statistics, and the rule of large numbers. "Are we alone?" said Neta Bahcall of Princeton University. "To me the answer is easy and obvious. Our sun is one star out of hundreds of billions of stars in our galaxy, and the Milky Way is just one out of billions and billions of galaxies. It's just impossible that we are the only life in the universe."

"I'm inclined to think that life is very common in the universe," said David Stevenson of Caltech. "I may turn out to be wrong, of course, but that's my working hypothesis."

In an interview conducted not long before his death, John Bahcall of Princeton said, "I am absolutely certain there is more life out there. It's one of the very few things for which I don't have any proof but on which I would bet a lot of money. The odds are so overwhelmingly in my favor."

Not only are there billions of stars, astronomers say, billions of solar ovens radiating photonic comestibles that practically beg to be eaten, but there are likely to be billions of planets circling those stars, billions of possible tables where one might find organisms that take in nutrients, excrete waste, replicate, and actually use the fondue set they got as a wedding present. Planetary formation, it seems, is a frequent byproduct of stellar condensation, the planetary disk forming as a result of the angular momentum of a collapsing, spinning star; anywhere from 10 to 50 percent of stars may have their share of circumstantial circumnavigators. Many astronomers are now searching for signs of extrasolar planets by checking for wobbles or irregularities in a star's motions, which may signal that it has gravitational companions, or the intermittent dimming of a star's light that would result whenever an orbiting planet passed between the star and us. And though for a while the only extrasolar planets astronomers could find were of the uninhabitable gas-giant category, more recently they have detected signs of smaller and possibly earthier worlds, tracing orbits at sensibly temperate distances from their parent suns.

Astronomers also find comfort in how relatively quickly life arose on Earth after the crust had cooled, and the unshakable conviction with which life has stood its ground ever since. They point to recent work in the field of nanotechnology, the chemistry of materials constructed on extremely small scales, showing that carbon molecules spontaneously form rings, tubes, and spheres, the very sort of skeletal structures on which life is draped. Carbon is a common constituent of supernova shrapnel, they say, and if carbon so readily self-assembles into the precursors of biomolecules, the rise of life may be virtually inevitable if carbon finds itself self-assembling in certain settings — for example, on a planet with liquid water to its credit. Again, not an outrageous demand. Water, like carbon, is commonplace, and though most of the cosmic quotient of water looks to be in gaseous or frozen form, there are sure to be other liquid oases in the vast sample space that is outer space. "Here on Earth, anywhere you find liquid water, you find life," said Andy Ingersoll of Caltech. "Life is remarkably robust when it comes to adapting to extremely cold or hot water, or very acidic water. It's hard to imagine, given the robustness of microbial life, that if there's liquid water somewhere else, life hasn't found a way to take advantage of it."

On the question of how complex any of that extraterrestrial life may be, and whether there are other technologically sophisticated civilizations with whom we in theory could communicate, astronomers become far more reserved. "When you start asking, what is the probability that life, once it has developed, will evolve something sufficiently intelligent that it tries to communicate and travel around, well, I don't think we're in a position to make a useful estimate of that," said Dave Stevenson of Caltech.

Nevertheless, a few resilient souls have sought to do exactly that.

Most famously, Frank Drake, then a Cornell astronomer and a founder of the Search for Extra-Terrestrial Intelligence initiative, or SETI, in the 1960s offered his methodical approach to calculating how many "communicative societies" may be out there in the Milky Way, a formulation now known as the Drake equation. Drake's take consists of seven variables to consider, proceeding from such comparatively straightforward factors as the rate of new star formation and the number of stars likely to have planets, and progressing into ever softer and more subjective terrain, including: the odds that a particular life-bearing locale will give rise to intelligent life; that the intelligence will be of a tinkering, toolmaking type; and, finally, that the technologically adroit civilization, having reached the point where it is capable of sending its halloos our way, will persist long enough to hear our reply.

Stevenson observes that the most uncertain and potentially deflating parameter of the Drake equation is the last one. "If the life span of an advanced civilization is only a few thousand years, then the probability of another intelligent civilization coexisting with us becomes low," said Stevenson. "Other civilizations might have come and gone before us, and new ones may be in the process of forming, but by the time they do we'll have destroyed ourselves. Either way, we could well be the only one in the galaxy at present."

But take heart! Remember that, while our naked night vision is limited mostly to the Milky Way, our sample space is not. Even if there were only one communicative society per galaxy, that still leaves us with billions of hypothetical entries on the Rolodex of hope. Admittedly, the terrible distances between galaxies could well preclude any communication beyond the science fictional, but it's good to think they're out there, those probabilistic star-flecked partners in space-time. And who knows? They may be better off than we are and have found the perfect intergalactic wormhole and are steadily heading our way. Please, please, stop by, any time, any stardate. We can't promise, but we will try, with all our heart and hemoglobin and every one of our 90 trillion body cells and our bacterial symbionts, too, to hang on, and dodge our own bullets, and be here when you arrive.

REFERENCES

ACKNOWLEDGMENTS

INDEX

References

1. Thinking Scientifically: An Out-of-Body Experience

Altschuler, Daniel R. *Children of the Stars: Our Origin, Evolution and Destiny.* Cambridge and New York: Cambridge University Press, 2002.

Atkins, Peter. *Galileo's Finger.* Oxford: Oxford University Press, 2003.

Ben-Shahar, Y., A. Robichon, M. B. Sokolowski, and G. E. Robinson. "Influence of Gene Action Across Different Time Scales on Behavior." *Science* 296 (2002): 741–44.

Bryson, Bill. *A Short History of Nearly Everything.* New York: Broadway Books, 2003.

Donald, Janet. *Learning to Think.* San Francisco: Jossey-Bass, 2002.

Eisner, Thomas. "Making the Microscope Loom Large in a Child's Life." *New York Times,* August 10, 2004. June 23, 2006 <www.nytimes.com>.

Emiliani, Cesare. *The Scientific Companion.* New York: Wiley, 1995.

Hazen, Robert M., and James Trefil. *Achieving Science Literacy.* New York: Doubleday, 1990.

Krauss, Lawrence M. *Fear of Physics.* New York: Basic Books, 1993.

Lustig, Cindy, Alex Konkel, and Larry L. Jacoby. "Which Route to Recovery?" *Psychological Science* 15 (2004): 729–35.

National Science Foundation. "Science and Engineering Labor Force." *Science and Engineering Indicators 2006.* September 6, 2006 <http://www.nsf.gov/statistics/seind06/c3/c3s2.htm#c3s212>.

Piel, Gerard. *The Age of Science.* New York: Basic Books, 2001.

Pollack, Henry N. *Uncertain Science . . . Uncertain World.* Cambridge: Cambridge University Press, 2003.

Remnick, David, ed. *Life Stories: Profiles from* The New Yorker. New York: Modern Library, 2001.

"Science Dull and Hard, Students Say." *BBC News.* September 6, 2006 <http://news.bbc.co.uk/1/hi/education/4100936.stm>.

Tallack, Peter, ed. *The Science Book.* London: Cassell, 2001.

Trefil, James. *The Nature of Science.* Boston: Houghton Mifflin, 2003.

Trefil, James, and Robert M. Hazen. *The Sciences: An Integrated Approach.* New York: Wiley, 2001.

Weinberg, Steven. "Can Science Explain Everything? Anything?" *The New York Review of Books.* May 31, 2001. January 31, 2002 <http://www.nybooks.com/articles>.

2. Probabilities: For Whom the Bell Curves

American Academy of Dermatology. "Melanoma Fact Sheet." *AAD Public Resource Center.* September 6, 2006 <http://www.aad.org/public/News/DermInfo/MelanomaFAQ.htm>.

Belkin, Lisa. "The Odds of That." *New York Times.* August 11, 2002.

Cohen, Jack, and Ian Stewart. "That's Amazing, Isn't It?" *New Scientist.* January 17, 1998.

Cohn, Victor. *News and Numbers.* Ames: Iowa State University Press, 1989.

Gonick, Larry, and Woollcott Smith. *The Cartoon Guide to Statistics.* New York: Harper Perennial, 1993.

"HIV Infection and AIDS." *National Institute of Allergy and Infectious Diseases.* September 6, 2006 <http://www.niaid.nih.gov/factsheets/hivinf.htm>.

Huff, Darrell. *How to Lie with Statistics.* New York: W. W. Norton, 1954.

Koehler, Jonathan J. "One in Millions, Billions, and Trillions." *Journal of Legal Education* 47 (1997): 214–23.

Kolata, Gina. "1-in-a-Trillion Coincidence, You Say? Not Really, Experts Find." *New York Times.* February 27, 1990.

Lee, Jennifer 8. "Who Needs Giacomo? Bet on the Fortune Cookie." *New York Times.* May 11, 2005.

Muller, Richard A. *Physics for Future Presidents.* Berkeley: University of California Press, 2004.

Palo Alto Medical Foundation. "The Darker Side of the Sun: Facts about Skin Cancer." September 6, 2006 <http://www.pamf.org/skincancer/>.

Paulos, John Allen. *Innumeracy.* New York: Vintage, 1990.

———. *A Mathematician Reads the Newspaper.* New York: Anchor, 1996.

Phillips, John L. *How to Think about Statistics.* New York: W. H. Freeman, 2000.

Pollack, Henry N. *Uncertain Science . . . Uncertain World.* Cambridge: Cambridge University Press, 2003.

Salsburg, David. *The Lady Tasting Tea: How Statistics Revolutionized Science in the Twentieth Century.* New York: W. H. Freeman, 2001.

Slovic, Paul. "Perception of Risk Posed by Extreme Events." *Risk Management Strategies in an Uncertain World.* Palisades, N.Y. April 12, 2002.

Taleb, Nassim N. "Learning to Expect the Unexpected." *Edge.* June 23, 2006 <http://www.edge.org/3rd_culture/taleb04/taleb_indexx.html>.

U.S. Department of Health and Human Services. "Results from the 2004 National Survey on Drug Use and Health: National Findings." *Office of Applied Studies.* <http://www.oas.samhsa.gov/NSDUH/2k4NSDUH/2k4results/2k4results.htm#ch4>.

Weiss, Rick. "Dazzled by 'Tortured Data.'" *Washington Post.* November 23, 1993. June 23, 2006 <http://www.nexis.com/research>.

Willett, Martin. "Bell Curves." *Debate Unlimited.* June 23, 2006 <http://mwillett.org/bell.htm>.

3. Calibration: Playing with Scales

American Society for Microbiology. "Monsters Among the Microbes." *Microbes.* September 1, 2005.

Ash, Russell. *The Top Ten of Everything 2004.* London and New York: Dorling Kindersley, 2003.

Calder, Nigel. *TimeScale.* New York: Viking, 1983.

Carpi, Anthony. "The Cell." *The Natural Sciences.* City University of New York. June 23, 2006 <http://web.jjay.cuny.edu/~acarpi/NSC/13-cells.htm>.

"Day." *Wikipedia.* September 6, 2006 <http://en.wikipedia.org/wiki/Day>.

Ford, Kenneth W. *The Quantum World.* Cambridge, Mass.: Harvard University Press, 2004.

Haldane, J.B.S. *On Being the Right Size.* Oxford: Oxford University Press, 1985.

Jaffe, Robert L. *The Time of Your Life — and Other Times.* Manuscript. 2005.

Lieberman, Abraham N. "What You Should Know about the Cell, DNA, and Genes." National Parkinson Foundation. October 6, 2005 <http://www.parkinson.org>.

Morrison, Philip, and Phylis Morrison. *Powers of Ten.* San Francisco: Scientific American Library, 1982.

NASA. "Solar History Timeline." *Solar-B.* September 6, 2006 <http://solarb.msfc.nasa.gov/science/timeline/index.html>.

"The Nervous System." *ThinkQuest.* September 6, 2006 <http://library.thinkquest.org/4371/About%20the%20Brain.htm>.

Pollock, Steven. *Particle Physics for Non-Physicists.* Chantilly, Va.: The Teaching Company, 2003.

Rensberger, Boyce. *Instant Biology.* New York: Fawcett Columbine, 1996.

Rigden, John S. *Hydrogen: The Essential Element.* Cambridge, Mass.: Harvard University Press, 2002.

Sandow, Stuart A., Chrissie Bamber, and J. W. Rioux. *Durations: The Encyclopedia of How Long Things Take.* New York: Times Books, 1977.

"Speed of a Bullet." *Science Education Partnerships.* August 1, 2003. Oregon State University, January 4, 2005 <http://www.seps.org/oracle/oracle.archive/Physical_Science.Physics>.

Sullivan, Jim. "How Big Is a . . . ?" *Cells Alive!* September 8, 2006 <http://www.cellsalive.com/howbig.htm>.

Tully, Brent. "How Big Is the Universe?" *NOVA Online.* University of Hawaii. June 23, 2006 <http://www.pbs.org/wgbh/nova/universe/howbig.html>.

"What Is the Average Speed of a Bullet Leaving the Muzzle of a Handgun?" *Answerbag.* January 4, 2005 <http://www.answerbag.com/>.

4. Physics: And Nothing's Plenty for Me

Altschuler, Daniel R. *Children of the Stars: Our Origin, Evolution and Destiny.* Cambridge and New York: Cambridge University Press, 2002.

Atkins, Peter. *Galileo's Finger.* Oxford: Oxford University Press, 2003.

Atkins, P. W. *The Periodic Kingdom.* New York: Basic Books, 1995.

Beaty, William J. "What Is 'Electricity'?" *Bill B's Science Hobbyist.* 1996. June 23, 2006 <http://www.amasci.com/miscon/whatis.html>.

Charap, John M. *Explaining the Universe.* Princeton: Princeton University Press, 2002.

Emsley, John. *Nature's Building Blocks.* Oxford: Oxford University Press, 2001.

European Space Agency. "Creation of Light Elements." *ESA High School Education.* April 9, 2003. December 1, 2005 <http://www.esa.int/esaED>.

Ferris, Timothy. *The Whole Shebang.* New York: Simon and Schuster, 1997.

Feynman, Richard P. *The Pleasure of Finding Things Out.* Cambridge, Mass.: Perseus, 1999.

Ford, Kenneth W. *The Quantum World.* Cambridge, Mass.: Harvard University Press, 2004.

Freudenrich, Craig. "How Light Works." *How Stuff Works.* June 23, 2006 <http://www.howstuffworks.com/light.htm>.

Gamow, George. *Mr Tompkins in Paperback.* Cambridge: Cambridge University Press, 1993.

Gleick, James. *Isaac Newton.* New York: Pantheon, 2003.

Gonick, Larry, and Art Huffman. *The Cartoon Guide to Physics.* New York: Harper Perennial, 1990.

Hazen, Robert M., and James Trefil. *Achieving Science Literacy.* New York: Doubleday, 1990.

Krauss, Lawrence M. *Fear of Physics.* New York: Basic Books, 1993.

Muller, Richard A. *Physics for Future Presidents.* Berkeley: University of California Press, 2004.

Murphy, Pat, and Paul Doherty. *The Color of Nature.* San Francisco: Chronicle, 1996.

NASA. "The Electromagnetic Spectrum." November 4, 2005 <http://imagers.gsfc.nasa.gov>.

Nave, Rod. "Quarks." *Hyperphysics.* George State University. September 8, 2006 <http://hyperphysics.phy-astr.gsu.edu/hbase/particles/quark.html#c6>.

"Observing Across the Spectrum." *Cool Cosmos: Multiwavelength Astronomy.* November 23, 2005 <http://www.coolcosmos.ipac.caltech.edu/cosmic_classroom>.

Overbye, Dennis. "The Universe Seems So Simple, Until You Have to Explain It." *New York Times.* October 22, 2002 <www.nytimes.com>.

Pollock, Steven. *Particle Physics for Non-Physicists.* Chantilly, Va.: The Teaching Company, 2003.

Rigden, John S. *Hydrogen: The Essential Element.* Cambridge, Mass.: Harvard University Press, 2002.

Schneider, Eric D., and Dorion Sagan. *Into the Cool: Energy Flow, Thermodynamics and Life.* Chicago: University of Chicago Press, 2005.

Senese, Fred. "Why Is Mercury a Liquid at STP?" *General Chemistry Online.* June 23, 2006 <http://antoine.frostburg.edu/chem/senese/101/periodic/faq/why-is-mercury-liquid.shtml>.

Trefil, James, and Robert M. Hazen. *The Sciences: An Integrated Approach.* New York: Wiley, 2001.

Weinberg, Steven. *Dreams of a Final Theory.* New York: Vintage, 1993.

Whittle, Mark. "A Brief History of Matter." *Prof. Mark Whittle's Home Page.* University of Virginia Department of Astronomy. September 8, 2006 <http://www.astro.virginia.edu/class/whittle/astr124/matter/matter_three.html>.

5. Chemistry: Fire, Ice, Spies, and Life

Angier, Natalie. "Free Radicals: The Price We Pay for Breathing." *New York Times.* April 25, 1993.

———. "Nonfinicky Vulture Wears Its Toxic Feast All Over Its Face." *New York Times.* April 30, 2002, Section F: 3.

——. "Serenade of Color Woos Pollinators to Flowers." *New York Times.* November 26, 1991, Section C: 4.

——. "Some Blend In, Others Dazzle: The Mysteries of Animal Colors." *New York Times.* July 20, 2004.

Ash, Russell. *The Top Ten of Everything 2004.* New York: Dorling Kindersley, 2003.

Atkins, Peter. *Galileo's Finger.* Oxford: Oxford University Press, 2003.

Atkins, P. W. *The Periodic Kingdom.* New York: Basic Books, 1995.

Ball, Philip. *Life's Matrix.* New York: Farrar, Straus and Giroux, 1999.

——. *Molecules: A Very Short Introduction.* Oxford: Oxford University Press, 2003.

Brain, Marshall. "How Food Works." *How Stuff Works.* June 23, 2006 <http://home.howstuffworks.com/food.htm>.

California Academy of Sciences. "Plants That Kill." *Science Now.* May 13, 2001. May 30, 2005 <http://www.calacademy.org/science_now/archive/wild_lives/california_carnivores_051301.htm>.

"Chinese Characters." *China Online.* September 6, 2006 <http://chineseculture.about.com/library/symbol/blcc_chemistry.htm>.

Coenders, A. *The Chemistry of Cooking.* Park Ridge, N.J.: Parthenon, 1992.

De Duve, Christian. *Vital Dust.* New York: Basic Books, 1995.

Eisner, Thomas. *For Love of Insects.* Cambridge, Mass.: Harvard University Press, 2003.

Emsley, John. *Nature's Building Blocks.* Oxford: Oxford University Press, 2001.

Garfield, Simon. *Mauve: How One Man Invented a Color That Changed the World.* New York: W. W. Norton, 2000.

Hoffmann, Roald. *The Same and Not the Same.* New York: Columbia University Press, 1995.

Hoffmann, Roald, and Vivian Torrence. *Chemistry Imagined.* Washington, D.C.: Smithsonian Institution, 1993.

Horgan, John. *The End of Science.* Reading, Mass.: Addison-Wesley, 1996.

McGovern, Patrick E., Juzhong Zhang, Jigen Tang, et al. "Fermented Beverages of Pre- and Proto-Historic China." *PNAS* 101 (2004): 17593–98.

"Molecular Structures." *Chemistry Guide.* April 6, 2005 <http://www.chemguide.co.uk/atoms/>.

Moore, John T. *Chemistry for Dummies.* New York: Wiley, 2003.

"Online Etymology Dictionary." September 6, 2006 <http://www.etymonline.com/index.php>.

Sacks, Oliver. *Uncle Tungsten: Memories of a Chemical Boyhood.* New York: Knopf, 2001.

6. Evolutionary Biology: The Theory of Every Body

Brumfiel, Geoff. "Who Has Designs on Your Students' Minds?" *Nature* 434 (2005): 1062–65.

Calder, Nigel. *TimeScale.* New York: Viking, 1983.

California Academy of Sciences. "Plants That Kill." *Science Now.* May 13, 2001. May 30, 2005 <http://www.calacademy.org/science_now/archive/wild_lives/california_carnivores_051301.htm>.

Campbell, Neil A., Lawrence G. Mitchell, and Jane B. Reece. *Biology: Concepts and Connections.* 3rd ed. San Francisco: Addison Wesley Longman, 2000.

Canadian Museum of Nature. "Star-Nosed Mole." *Nature.Ca.* April 29, 2005 <http://www.nature.ca>.

"Ceratobatrachus Guentheri; Solomons Leaf Frog." *Digital Library Project.* Berkeley: University of California. May 10, 2005 <http://elib.cs.berkeley.edu>.

Cornish, Jim. "Penguins: General Information." *General Resources.* Classroom Connect. May 18, 2006 <http://www.cdli.ca/CITE/penguins_general.htm>.

Cracid Specialist Group. "What Is a Cracid?" May 18, 2005 <http://www.cracids.org/what_is_a_cracid.html>.

Dean, Cornelia. "Challenged by Creationists, Museums Answer Back." *New York Times.* September 20, 2005.

De Duve, Christian. *Vital Dust.* New York: Basic Books, 1995.

Delacour, Jean. *Curassows and Related Birds.* New York: American Museum of Natural History, 1973.

Delong, Edward F. "A Plentitude of Ocean Life." *Natural History.* May 2003.

"The Duck-Billed Platypus." *The Duck-Billed Platypus.* April 29, 2005 <http://www.genevaschools.org>.

Ehrlich, Paul R. *Human Natures.* Washington, D.C.: Island Press, 2000.

Eisner, Thomas. *For Love of Insects.* Cambridge, Mass.: Harvard University Press, 2003.

Fortey, Richard. *Life: A Natural History of the First Four Billion Years of Life on Earth.* New York: Vintage, 1997.

Fortey, Richard A. *Trilobite!* London: HarperCollins, 2000.

"General Characteristics of Primates." *The Primates: Overview.* April 1, 2005. May 19, 2005 <http://anthro.palomar.edu/primate/prim_1.htm>.

Gore, Pamela J. "The PreCambrian." *Georgia Perimeter College.* June 24, 2006 <http://www.gpc.edu/~pgore/geology/geo102/precamb.htm>.

"Hallucigenia." *Answers.Com.* September 6, 2006 <http://www.answers.com/topic/hallucigenia>.

Harris, Paul. "Mixing Science with Creationism." *Salon.* May 24, 2005 <http://www.salon.com/news/feature>.

Keller, Bill. "God and George W. Bush." *New York Times.* May 17, 2003, Section A: 17.

Knoll, Andrew H. *Life on a Young Planet.* Princeton: Princeton University Press, 2003.

Knoll, Andrew H., and Sean B. Carroll. "Early Animal Evolution: Emerging Views from Comparative Biology and Geology." *Ecology* 284 (1999): 2129–37. June 24, 2006 <http://cas.bellarmine.edu/tietjen/Ecology/early_animal_evolution.htm>.

Miller, Kenneth R. *Finding Darwin's God.* New York: Cliff Street Books, 1999.

Miller, Kenneth R., and Joseph S. Levine. *Biology: Discovering Life.* 2nd ed. Lexington, Mass.: D. C. Heath, 1994.

Minkoff, Eli C., and Pamela J. Baker. *Biology Today.* 2nd ed. New York: Garland, 2001.

Patuxent Bird Population Studies. "Painted Bunting Passerina Ciris." *Painted Bunting Identification Tips.* April 29, 2005 <http://www.mbr-pwrc.usgs.gov>.

Quammen, David. "Was Darwin Wrong?" *National Geographic.* November 2004: 4–31.

Raven, Peter, et al. *Biology.* 7th ed. Boston: McGraw Hill, 2005.

Saletan, William. "Creationism Evolves." *Slate.* September 1, 1999. May 25, 2005 <http://www.slate.com>.

——. “What Matters in Kansas.” *Slate.* May 11, 2005. May 25, 2005 <http://www.slate.com>.

Sever, Megan. “Creationism in a National Park.” *Geotimes.* March 2004. May 5, 2005 <http://www.geotimes.org/mar04/NN_grandcanyoncreation.html>.

Smithsonian Institution. “Genus: Hallucigenia Sparsa.” *National Museum of Natural History Department of Paleobiology.* September 6, 2006 <http://www.nmnh.si.edu/paleo/shale/phallu.htm>.

Southwood, Richard. *The Story of Life.* Oxford: Oxford University Press, 2003.

Sze, Emily Lei Pi. “Theodosius Dobzhansky.” May 16, 2005 <http://www.mnsu.edu/emuseum/information/biography>.

“Tiger Swallow Butterfly.” *Enchanted Learning.* September 6, 2006 <http://www.enchantedlearning.com/subjects/butterfly/species/Tigersw.shtml>.

Tobin, Allan J., and Jennie Dusheck. *Asking About Life.* 2nd ed. Orlando and Philadelphia: Harcourt, 2001.

University of Manitoba. “Star-Nosed Mole.” *The Mole Tunnel.* April 29, 2005 <http://home.cc.umanitoba.ca>.

University of Michigan Museum of Zoology. “Family Hominidae.” *Animal Diversity Web.* September 6, 2006 <http://animaldiversity.ummz.umich.edu/site/accounts/information/Hominidae.html>.

Weinberg, Steven. “Can Science Explain Everything? Anything?” *New York Review of Books.* May 31, 2001. January 31, 2002 <http://www.nybooks.com/articles>.

7. Molecular Biology: Cells and Whistles

American Society for Microbiology. “Monsters Among the Microbes.” *Microbes.* September 1, 2005 <http://www.microbe.org/microbes/biggest.asp>.

Angier, Natalie. “Free Radicals: The Price We Pay for Breathing.” *New York Times.* April 25, 1993.

Atkins, Peter. *Galileo’s Finger.* Oxford: Oxford University Press, 2003.

Atkins, P. W. *The Periodic Kingdom.* New York: Basic Books, 1995.

“Bacteria in the Human Body.” *Wikipedia.* August 24, 2005 <http://en.wikipedia.org/wiki/Bacteria_in_the_human_body>.

Ball, Philip. *Life’s Matrix.* New York: Farrar, Straus and Giroux, 1999.

Carey, Bjorn. “Wild Things: The Most Extreme Creatures.” *Live Science Animal World.* February 7, 2005. August 26, 2005 <http://www.livescience.com/animalworld/050207_extremophiles.html>.

Carpi, Anthony. “The Cell.” *The Natural Sciences.* City University of New York. June 23, 2006 <http://web.jjay.cuny.edu/~acarpi/NSC/13-cells.htm>.

“The Cell: Down to Basics.” *Beyond Books: Life Science.* September 1, 2005 <http://www.beyondbooks.com>.

“Cells.” *Biosciences: Science About Life.* September 1, 2005 <http://www.saasta.ac.za/ biosciences/cells.html>.

Conniff, Richard. “Body Beasts.” *National Geographic.* December 1998. August 24, 2005 <http://www.nationalgeographic.com/ngm/9812/fngm/>.

“The DNA Codons.” *The Genetic Code.* September 21, 2005 <http://users.rcn.com/jkimball.ma.ultranet/BiologyPages/C/Codons.html>.

Emsley, John. *Nature’s Building Blocks.* Oxford: Oxford University Press, 2001.

Lieberman, Abraham N. “What You Should Know about the Cell, DNA, and

Genes." National Parkinson Foundation. October 6, 2005 <http://www.parkinson.org>.

Madanecki, Piotr. "Luminescent Bacteria." August 26, 2005 <http://www.biology.pl/bakterie_sw>.

"My Favorite Protein: Insulin." *My Favorite Protein.* Davidson College. September 28, 2005 <http://www.bio.davidson.edu/Courses/Mobio>.

"Questions and Answers about Biology." *Ken Miller and Joe Levine.* October 7, 2005 <http://www.millerandlevine.com/ques/eggs.html>.

Raven, Peter, et al. *Biology.* 7th ed. Boston: McGraw Hill, 2005.

Rensberger, Boyce. *Instant Biology.* New York: Fawcett Columbine, 1996.

Rigden, John S. *Hydrogen: The Essential Element.* Cambridge, Mass.: Harvard University Press, 2002.

Sullivan, Jim. "How Big Is a . . . ?" *Cells Alive!* September 8, 2006 <http://www.cellsalive.com/howbig.htm>.

"The Theory of Differential Gene Expression." *Developmental Genetics.* October 6, 2005 <http://www.emunix.emich.edu>.

8. Geology: Imagining World Pieces

"Aerobic/Anaerobic Systems." *Book Rags Biology Study Guide.* February 10, 2006 <http://www.BookRags.com>.

Altschuler, Daniel R. *Children of the Stars: Our Origin, Evolution and Destiny.* Cambridge and New York: Cambridge University Press, 2002.

American Society for Microbiology. "Monsters Among the Microbes." *Microbes.* September 1, 2005 <http://www.microbe.org/microbes/biggest.asp>.

"Asteroids, Comets, and Meteoroids." *BBC-H2g2.* November 4, 2005 <www.bbc.co.uk/dna/h2g2>.

Bercovici, David, Yanick Ricard, and Mark A. Richards. "The Relation Between Mantle Dynamics and Plate Tectonics: A Primer." *Geophysical Monograph* 121 (2000): 5–46.

Buffett, Bruce A. "Geophysics: The Thermal State of Earth's Core." *Science* 299 (2003): 1675–77.

Calder, Nigel. *TimeScale.* New York: Viking Press, 1983.

"The Changing Earth and Cyanobacteria: The Oxygen Revolution." *Carleton Museum.* June 24, 2006 <www.carleton.ca/Museum/stromatolites/OXYGEN.htm>.

"Clostridium." *Medic.Uth.* February 10, 2006 <http://medic.uth.tmc.edu/path>.

Darling, David. "Ocean's Origin." *The Worlds of David Darling.* June 23, 2006 <http://www.daviddarling.info/encyclopedia/O/oceansorigin.html>.

"The Different Plates; Sea Floor Spreading; Subduction." *ThinkQuest.* January 26, 2006.

"Earth's Structure." *ThinkQuest Team.* February 17, 2006 <http://mediatheek.thinkquest.nl>.

"Feeding in Green Plants." *The Open Door Web Site.* May 4, 2005. May 30, 2005 <http://www.saburchill.com/chapters/chap0027.html>.

Fortey, Richard. *Life: A Natural History of the First Four Billion Years of Life on Earth.* New York: Vintage, 1997.

Fortey, Richard A. *Trilobite!* London: HarperCollins, 2000.

Fountain, Henry. "When Giants Had Wings and 6 Legs." *New York Times.* February 3, 2004.

"Geothermal Energy." *Kansas Energy Education Foundation.* June 23, 2006.

Gilman, Larry, and K. Lee Lerner. "Ocean-Floor Bathymetry." *Water Encyclopedia.* September 14, 2006 <http://www.waterencyclopedia.com/Oc-Po/Ocean-Floor-Bathymetry.html>.

Gore, Pamela J. "The PreCambrian." *Georgia Perimeter College.* June 24, 2006 <http://www.gpc.edu/~pgore/geology/geo102/precamb.htm>.

Hinshaw, Dorothy P. *Shaping the Earth.* New York: Clarion, 2000.

Jeffares, Daniel C., and Anthony M. Poole. "Were Bacteria the First Forms of Life on Earth?" *ActionBioscience.* December 2000. American Institute of Biological Sciences. October 10, 2005 <http://www.ActionBioscience.org>.

"Journey to the Center of the Earth Synopsis." *Wikipedia.* February 22, 2006 <http://en.wikipedia.org/wiki/Journey_to_the_Center_of_the_Earth>.

Kandel, Robert. *Water from Heaven.* New York: Columbia University Press, 2003.

Knoll, Andrew H. *Life on a Young Planet.* Princeton: Princeton University Press, 2003.

Knoll, Andrew H., and Sean B. Carroll. "Early Animal Evolution: Emerging Views from Comparative Biology and Geology." *Ecology* 284 (1999): 2129–37. June 24, 2006 <http://cas.bellarmine.edu/tietjen/Ecology/early_animal_evolution.htm>.

Levy, Sharon. "Navigating with a Built-in Compass." *National Wildlife Magazine.* October–November 1999. National Wildlife Federation. October 10, 2005 <http://www.org/nationalwildlife>.

Louie, J. "Earth's Interior." *Seismological Laboratory.* October 10, 1996. University of Nevada, Reno. January 19, 2006 <http://www.seismo.unr.edu>.

Martin, William, and Miklos Mueller. "The Hydrogen Hypothesis for the First Eukaryote." *Nature* 392 (1998): 37–41.

Mathez, Edmond A., ed. *Earth: Inside and Out.* New York: American Museum of Natural History, 2001.

Minarik, William. "The Multi-Anvil Press at Work." *Studying the Earth's Formation.* September 18, 2006 <http://www.llnl.gov/str/Minarik.html>.

Monastersky, Richard. "Ancient Animals Got a Rise Out of Oxygen." *Science News.* May 13, 1995.

Moores, Eldridge, ed. *Shaping the Earth: Tectonics of Continents and Oceans.* New York: W. H. Freeman, 1990.

NASA. "Evidence Supporting Continental Drift." *NASA: On the Move — Continental Drift and Plate Tectonics.* January 10, 2006 <http://www.earth.nasa.gov>.

NASA Goddard Spaceflight Center. "The Water Cycle." September 14, 2006 <http://neptune.gsfc.nasa.gov/education/pdf/Water_Cycle_Litho.pdf#search=%22total%20water%20earth%20nasa%20326%20trillion%22>.

Pendick, Daniel. "Earth: All Stressed Out." *Savage Earth.* PBS. January 5, 2006 <http://www.pbs.org/wnet/savageearth>.

Robertson, Eugene C. "The Interior of the Earth." *USGS.* June 23, 2006 <http://pubs.usgs.gov/gip/interior/>.

"Structure of the Earth." *Fundamentals of Physical Geography.* February 17, 2006 <http://www.physicalgeography.net/fundamentals>.

Svitil, Kathy. "The Earth at Work." *Savage Earth.* PBS. January 5, 2006 <http://www.pbs.org/wnet/savageearth>.

UK National HPC Service. University of Manchester. "Turing Probes the Earth's Core." September 14, 2006 <http://www.csar.cfs.ac.uk/about/csarfocus/focus4/core.pdf#search=%22pressures%20Earth's%20core%22>.

University College, London. "The Development of Life on Earth." *UCL Diploma Course.* August 26, 2005 <http://www.star.ucl.ac.uk>.

"USGS Science for a Changing World." *USGS Publications.* September 14, 2006 <http://pubs.usgs.gov/gip/>.

Vaiden, Robert C. "Plate Tectonics: Mysteries Solved!" *ISGS Geobit 10.* Illinois State Geological Survey. January 26, 2006 <http://www.isgs.uiuc.edu/servs/pubs/geobits-pub>.

Valley, John W. "A Cool Early Earth?" Scientificamerican.com. September 26, 2005. February 10, 2006 <www.sciam.com>.

Verne, Jules. "A Journey to the Center of the Earth Chapter XXXVI." *Jules Verne Collection.* February 23, 2006 <http://jv.gilead.org.il/vt/c_earth/36.html>.

Washington State University. "The Big Questions — Photosynthesis." *Ask Dr. Universe.* February 13, 2006 <http://www.wsu.edu/DrUniverse/>.

"What Do We Know about the Origins of the Earth's Oceans?" *Scientific American.* June 23, 2006 <http://www.sciam.com/askexpert_question.cfm?articleID=00085119-C6F1-1C71-9EB7809EC588F2D7&catID=3&topicID=22>.

Winchester, Simon. *A Crack in the Edge of the World.* New York: HarperCollins, 2005.

9. Astronomy: Heavenly Creatures

Altschuler, Daniel R. *Children of the Stars: Our Origin, Evolution and Destiny.* Cambridge and New York: Cambridge University Press, 2002.

"Asteroids, Comets, and Meteoroids." *BBC-H2g2.* November 4, 2005 <www.bbc.co.uk/dna/h2g2>.

"Beginnings: Space and Time." *Science and Nature: Space.* October 7, 2005 <bbc.co.uk>.

"Big Bang." *Wikipedia.* November 10, 2005 <http://en.wikipedia.org/wiki/Big_Bang>.

"Big Bang Theory." *Creation of a Cosmology.* November 23, 2005 <http://ssscott.tripod.com/BigBang.html>.

"Biography of a Star: Our Sun's Birth, Life, and Death." *The Universe in a Classroom.* December 20, 2005 <http://www.astrosociety.org/education/publications>.

"Birth of Stars and Galaxies." *PBS: Mysteries of Deep Space.* November 10, 2005 <www.pbs.org>.

"Blast from the Past: Farthest Supernova Ever Seen Sheds Light on Dark Universe." *HubbleSite.* April 2, 2001. November 10, 2005 <http://hubblesite.org/newscenter/newsdesk/archive>.

Boughn, Stephen, and Robert Crittenden. "A Correlation Between the Cosmic Microwave Background and Large-Scale Structure in the Universe." *Nature* 427 (2004): 45–48.

Britt, Robert R. "Freeze, Fry or Dry: How Long Has the Earth Got?" *Space.* February 25, 2000. November 24, 2005 <http://www.space.com>.

———. "Most Distant Galaxy Hints at Dark Ages." *Space.* February 16, 2004. November 10, 2005 <http://www.space.com/scienceastronomy>.

——. "Our Tiny Universe: What's Really Visible at Night." *Space.* December 29, 2003. November 20, 2005 <http://www.space.com>.

——. "The Reality of Antimatter." *Space.* September 29, 2003. November 28, 2005 <http://www.space.com>.

Chang, Kenneth. "Dying Star Flares Up, Briefly Outshining Rest of Galaxy." *New York Times.* February 20, 2005, Section 1: 26.

——. "Tiny, Plentiful and Really Hard to Catch." *New York Times.* April 26, 2005.

Charap, John M. *Explaining the Universe.* Princeton, N.J.: Princeton University Press, 2002.

"Colonization of the Outer Solar System." *Wikipedia.* November 23, 2005 <http://en.wikipedia.org/wiki/Colonization_of_the_outer_solar_system>.

"Discovery of the Cosmic Background Radiation." *Footprints of Creation.* November 21, 2005 <http://archive.ncsa.uiuc.edu>.

European Space Agency. "Creation of Light Elements." *ESA High School Education.* April 9, 2003. December 1, 2005 <http://www.esa.int/esaED>.

"Evolution of Stars." *The Milky Way.* December 20, 2005 <http://www.milky-way.com/gb/sevol.htm>.

Ferris, Timothy. *The Whole Shebang.* New York: Simon and Schuster, 1997.

"Formation of the Solar System." *Search for Planets.* European Space Agency. December 19, 2005 <http://sci2.esa.int/interactive/media/>.

Freudenrich, Craig. "How Light Works." *How Stuff Works.* June 23, 2006 <http://www.howstuffworks.com/light.htm>.

Grinspoon, David. *Lonely Planets.* New York: Ecco, 2003.

Hakim, Joy. *The Story of Science.* Washington, D.C.: Smithsonian Books, 2004.

Hamilton, Calvin J. "Star Formation, Life, and Death." *View of the Solar System.* December 20, 2005 <http://www.solarviews.com/eng/starformation.htm>.

Hawley, John F. "Is There Extraterrestrial Life in the Universe?" *John F. Hawley.* University of Virginia. November 23, 2005 <http://www.astro.virginia.edu>.

Hazen, Robert M., and Maxine Singer. *Why Aren't Black Holes Black?* New York: Anchor Books, 1997.

Kong, Patricia. "Speed of the Milky Way in Space." *The Physics Factbook.* November 18, 2005 <http://hypertextbook.com/facts>.

LaRocco, Chris, and Blair Rothstein. "The Big Bang." *Chris LaRocco and Blair Rothstein Present.* University of Michigan. Nov. 10, 2005 <http://www.umich.edu/>.

"Milky Way." *Wikipedia.* November 18, 2005 <http://en.wikipedia.org/wiki/Milky_Way>.

Monterey Institute for Research in Astronomy. "Why Do Stars Twinkle?" *MIRA.* February 16, 1999. May 10, 2005 <http://www.mira.org>.

NASA. "Discovery of the Cosmic Microwave Background." *WMAP Cosmology 101.* November 21, 2005 <http://map.gsfc.nasa.gov>.

——. "The Electromagnetic Spectrum." November 4, 2005 <http://imagers.gsfc.nasa.gov>.

——. "How Did the Solar System Form?" *Science@Nasa.* November 28, 2005. December 19, 2005 <http://science.hg.nasa.gov/solar_system/science/formation.html>.

——. "Nucleosynthesis." *Cosmicopia.* December 21, 2005 <http://helios.gsfc.nasa.gov>.

——. "Solar History Timeline." *Solar-B.* September 6, 2006. <http://solarb.msfc.nasa.gov/science/timeline/index.html>.

"Observing Across the Spectrum." *Cool Cosmos: Multiwavelength Astronomy.* November 23, 2005 <http://www.coolcosmos.ipac.caltech.edu/cosmic_classroom>.

"The Odds Against ET." *Popular Mechanics.* November 1, 2000. October 10, 2005 <http://www.popularmechanics.com/science/space/1282586.html>.

"Odds of Complex Life: Great Debates." *Astrobiology Magazine.* October 19, 2005 <http://www.astrobio.net/news>.

"Origins of the Days of the Week." *Aerospaceweb.* October 28, 2005 <http://www.aerospaceweb.org/question/astronomy>.

Overbye, Dennis. "The Universe Seems So Simple, Until You Have to Explain It." *New York Times.* October 22, 2002 <www.nytimes.com>.

"Physical Environment: Red Giant." December 20, 2005 <http://www.historyoftheuniverse.com/starold.html>.

Pine, Ronald C. "Introduction: Our Cosmological Roots." *Science and the Human Prospect.* University of Hawaii. June 23, 2006 <http://www.hcc.hawaii.edu/~pine/book1qts/chapter1qts.htm>.

Preuss, Paul. "A Supernova Named Albinoni Is the Oldest and Farthest Ever Found." *Lawrence Berkeley National Lab.* December 17, 1998. November 10, 2005 <http://www.lbl.gov/supernova/albinoni.html>.

"Quasars." *Raindrop Laboratories.* November 4, 2005 <http://www.rdrop.com/users/green/school>.

"Redshift." *Wikipedia.* November 10, 2005 <http://en.wikipedia.org/wiki/Redshift>.

"The Seven-Day Week and the Meaning of the Names of the Days." October 28, 2005 <http://www.crowl.org/Lawrence/time/days.html>.

"The Shape of the Milky Way." November 4, 2005 <http://homepage.mac.com/rarendt/Galaxy/mw.html>.

Shostak, Seth. "The Holy Grail: Small, Rocky Worlds." *SETI Institute.* February 2, 2006. February 10, 2006 <http://www.seti.org>.

Singh, Simon. *Big Bang.* New York: HarperCollins, 2004.

Sloan Digital Sky Survey/SkyServer. "The Expanding Universe." *Astronomy.* November 20, 2005 <http://cas.sdss.org>.

"Stars." *BBC-H2g2.* November 4, 2005 <www.bbc.co.uk/dna/h2g2>.

"Stellar Nucleosynthesis." *Lives and Deaths of Stars.* December 1, 2005 <http://www.astronomynotes.com>.

"Sun, the Solar System's Only Star." *Astronomy Today.* September 6, 2006 <http://www.astronomytoday.com/astronomy/sun.html>.

"The 305 Meter Radio Telescope." *National Astronomy and Ionosphere Center Arecibo Observatory.* November 16, 2005 <http://www.naic.edu/public>.

Tully, Brent. "How Big Is the Universe?" *NOVA Online.* University of Hawaii. June 23, 2006 <http://www.pbs.org/wgbh/nova/universe/howbig.html>.

University Corporation for Atmospheric Research. "Solar System Formation." *Windows to the Universe.* December 19, 2005 <http://www.windows.ucar.edu>.

Webster, Guy. "Howdy, Strangers." *Jet Propulsion Laboratory.* August 19, 2002. June 24, 2006 <http://www.jpl.nasa.gov/news/features.cfm?feature=555>.

Weinberg, Steven. "Can Science Explain Everything? Anything?" *New York Review of Books.* May 31, 2001. January 31, 2002 <http://www.nybooks.com/articles>.

——. *Dreams of a Final Theory.* New York: Vintage, 1993.

Whittle, Mark. "A Brief History of Matter." *Prof. Mark Whittle's Home Page.* University of Virginia Department of Astronomy. September 8, 2006 <http://www.astro.virginia.edu/class/whittle/astr124/matter/matter_three.html>.

Yarris, Lynn. "Discovery of Most Distant Supernovas." *Discovery of Distant Supernovas.* January 16, 1996. Lawrence Berkeley Lab. November 10, 2005 <http://www-supernova.lbl.gov>.

Acknowledgments

I know I'm not the only writer who, at some point halfway through a major book project, devotes several days to a single, urgent task: searching for the perfect excuse to give up. A persuasive, shame-retardant excuse that would allow me to return the advance, recycle the spiral notebooks, and reformat the hard drive without also having to abandon my family and move to Niles, Michigan.

As with midpoint panics past, however, I could find no graceful way out. I couldn't bear to disappoint my daughter ("Book? You were writing a book?") and I really couldn't tolerate the idea of having wasted the time of so many smart, generous, overtaxed people.

I therefore would like to thank first and most fervently all the scientists who contributed to this book. I thank them for their in-depth knowledge, their succinct explanations of the most daunting concepts, their blue-sky but always scrupulous musings, their colorful comparisons, and their collegial wisecracks. Even the many researchers whom I interviewed but did not end up quoting by name informed the narrative in letter and spirit alike. *The Canon* obviously would not exist without the legions of scientific minds from which I freely plundered. My gratitude to them knows no bounds.

I owe particular thanks to the scientists who reviewed parts of the manuscript prior to publication: Jackie Barton, Brian Greene, Alan Guth, Jo Handelsman, Kip Hodges, Jonathan Koehler, Gene Robinson, Donald Sadoway, Meg Urry, and David Wake. They set me straight in places where I'd gone factually, logically, conceptually, or aesthetically astray; they suggested crucial additions and sensible subtractions; and they generally spared me the lasting consequences of multiple self-inflicted wounds.

In addition to their expert input, the book was fact-checked line by line as thoroughly as possible. Nevertheless, flubs and fatuities undoubtedly remain, for which I alone should be flogged.

I must also thank the many yeomanly press officers at the universities I visited, for their help in arranging and scheduling interviews, rearranging and rescheduling interviews, soothing professorial tempers when one prolonged interview made me late for the next, and even, during a particularly harried trip, setting my daughter up with a local day camp so that I could work uninterrupted.

Thanks to my editors, Amanda and Jayne, and my agent, Anne, for helping to hammer the book into shape, and being patient, and laughing at my jokes, and making me feel just a little less lost and alone. But because writing requires a clausura, a calm and solitary space, I am indebted to Bruce Martin and the Library of Congress for granting me a writer's office where I found focus and grace.

Thanks to Dennis for his cosmic expertise, and to Nancy for not once asking me "How's the book coming along?"

And then there's Rick. My trying to express my gratitude here is a bit like the citizens of Heorot, once Beowulf had slain their nemesis, Grendel, saying thank you with a gift card for Starbucks. Rick is a fellow writer and lover of science, and he contributed immeasurably to every phase of the project. He interviewed scientists and he helped shape the book's structure and identify its major and minor themes. He gathered string, ropes, lifelines. He sacrificed nights, weekends, and vacation days to the enterprise, but he never sacrificed the rigor of his judgment or the clarity of his mind. Time and again he beat back my demons and Grendels. Good thing I make decent coffee at home.

Index